Hydrography of and Biogeochemical Inputs to Liberty Bay, a Small Urban Embayment in Puget Sound, Washington

Edited by Renee K. Takesue

A pilot study by the Effects of Urbanization Task of the U.S. Geological Survey Multi-Disciplinary Coastal Habitats in Puget Sound Project

Scientific Investigations Report 2011–5152

U.S. Department of the Interior
U.S. Geological Survey

U.S. Department of the Interior
KEN SALAZAR, Secretary

U.S. Geological Survey
Marcia K. McNutt, Director

U.S. Geological Survey, Reston, Virginia: 2011

For more information on the USGS—the Federal source for science about the Earth, its natural and living resources, natural hazards, and the environment, visit http://www.usgs.gov or call 1–888–ASK–USGS.

For an overview of USGS information products, including maps, imagery, and publications, visit http://www.usgs.gov/pubprod

To order this and other USGS information products, visit http://store.usgs.gov

Suggested citation:
Takesue, Renee K., ed., 2011, Hydrography of and biogeochemical inputs to Liberty Bay, a small urban embayment in Puget Sound, Washington: U.S. Geological Survey Scientific Investigations Report 2011–5152, 98 p.

Preface

This multi-chapter report describes scientific and logistic understanding gained from a 2 year proof-of-concept study in Liberty Bay, a small urban embayment in central Puget Sound, Washington. The introductory chapter describes the regional and local setting, the high-level study goals, the site-specific urban stressors, and the interdisciplinary study approach. Subsequent data chapters describe detailed studies of various components of the Liberty Bay ecosystem: the aquatic environment (Chapter 2), surface and groundwater quantity and quality (Chapter 3), sediment quality (Chapter 4), eelgrass habitat (Chapter 5), carbon and nitrogen sources (Chapter 6), and a statistical model relating herring spawn probability to shoreline attributes (Chapter 7). The final chapter synthesizes knowledge about individual components into a system-wide understanding of how urbanization may affect the Liberty Bay ecosystem. The Liberty Bay study was conducted as part of the U.S. Geological Survey's Coastal Habitats in Puget Sound project, an interdisciplinary collaboration to understand physical and biological processes that affect nearshore ecosystems.

Contents

Conversion Factors, Datums, and Abbreviations and Acronyms

Conversions Factors

Inch/Pound to SI

Multiply	By	To obtain
Length		
mile (mi)	1.609	kilometer (km)

SI to Inch/Pound

Multiply	By	To obtain
Length		
centimeter (cm)	0.3937	inch (in)
meter (m)	3.281	foot (ft)
kilometer (km)	0.6214	mile (mi)
Area		
hectare (ha)	2.471	acre
square kilometer (km^2)	247.1	acre
hectare (ha)	0.003861	square mile (mi^2)
square kilometer (km^2)	0.3861	square mile (mi^2)
Flow rate		
cubic meter per second (m^3/s)	70.07	acre-foot per day (acre-ft/d)
meter per second (m/s)	3.281	foot per second (ft/s)
cubic per second (m^3/s)	35.31	cubic foot per second (ft^3/s)
Mass		
gram (g)	0.03527	ounce, avoirdupois (oz)

Temperature in degrees Celsius (°C) may be converted to degrees Fahrenheit (°F) as follows:

$$°F=(1.8×°C)+32.$$

Concentrations of chemical constituents in water are given in milligrams per liter (mg/L), micrograms per liter (µg/L), or moles per liter (mol/L).

Datums

Vertical coordinate information is referenced to the North American Vertical Datum of 1988 (NAVD 88).

Horizontal coordinate information is referenced to the North American Datum of 1927 (NAD 27).

Altitude, as used in this report, refers to distance above the vertical datum.

Conversion Factors, Datums, and Abbreviations and Acronyms—Continued

Abbreviations and Acronyms

ADCP	acoustic Doppler current profiler
C	carbon
CHIPS	Coastal Habitats in Puget Sound Project
CI	confidence interval
CT	conductivity, temperature
CTD	conductivity, temperature and depth
DIN	dissolved inorganic nitrogen
DO	dissolved oxygen
DOC	dissolved organic carbon
EM	electromagnetic
HCl	Hydrochloric acid
HT	high tide
ICP-MS	Inductively-coupled plasma-mass spectrometry
LB	Liberty Bay mooring site
LT	low tide
M	moles per liter
MLLW	mean lower-low water
N	nitrogen
NTU	nephelometric turbidity unit
NWQL	National Water Quality Laboratory (USGS)
OBS	optical backscatter sensor
ORP	oxidation-reduction potential
PAHs	polycyclic aromatic hydrocarbons
PB	Point Bolin mooring site
PCBs	polychlorinated biphenyls
POC	particulate organic carbon
POCIS	Polar Organic Chemical Integrative Sampler
POM	particulate organic matter
PPCPs	pharmaceutical and personal care products
PSAMP	Puget Sound Ambient Monitoring Program
psu	practical salinity unit
SBE	Seabird Electronics
SE	standard error
SGD	submarine groundwater discharge
SSC	suspended-sediment concentration
TSS	total suspended solids
Urban CHIPS	Effects of Urbanization Task of USGS CHIPS
USGS	U.S. Geological Survey
V	volt
VOC	volatile organic compound
WWTP	wastewater treatment plant

Chapter 1. Overview of Effects of Urbanization on the Nearshore Ecosystem of Puget Sound, Washington

By Renee K. Takesue[1], Richard S. Dinicola[2], Jessica R. Lacy[1], Theresa L. Liedtke[3], Dennis W. Rondorf[3], Collin D. Smith[3], and Raymond D. Watts[4]

Background

Puget Sound is the second largest estuary in the United States with more than 2,000 mi of shoreline (fig. 1-1). The estuary supports a diverse and productive ecosystem with cultural, commercial, and recreational value for residents and visitors. The natural beauty of Puget Sound makes it a desirable place to live; its natural resources are used for commercial gain, and its deep, glacier-carved channels are ideal locations for commercial port facilities and U.S. Naval installations. During the past 150 years, Puget Sound has experienced rapid residential, urban, and industrial development that is expected to continue. However, this development in the coastal watersheds of Puget Sound has destroyed or altered nearshore habitats. In Puget Sound, 75 percent of the salt marshes are lost, 33 percent of the shoreline is altered, and significant declines are reported for more than 40 species of concern including orcas, bald eagles, and salmon that depend on the nearshore for food, shelter, or spawning. The impairment of nearshore processes is a critical factor in the declining health of Puget Sound (MacDonald and others, 1994; Thom and others, 1994).

The U.S. Geological Survey (USGS) Multi-Disciplinary Coastal Habitats in Puget Sound Project (CHIPS) addresses the need for scientific understanding about the physical and biogeochemical processes and socio-economic values that shape and sustain healthy nearshore habitats and ecosystems. Non-natural disturbances, or impacts, affect the nearshore ecosystems by altering factors that shape habitat structures and processes (fig. 1-2).

The Effects of Urbanization Task of USGS CHIPS (Urban CHIPS) investigates how non-natural disturbances associated with urban development and human activities in coastal watersheds and along shorelines alter physical, chemical, and biological conditions and processes in the nearshore. A conceptual model of urban disturbances in the coastal zone provides a framework for investigating effects of urbanization on nearshore ecosystems (fig. 1-3). Disturbances associated with urbanization can be divided into two categories based on how the disturbance is delivered to the nearshore: from urban watersheds or from direct modifications of the shore (fig. 1-3). Urban development in coastal watersheds may affect the timing and magnitude of freshwater, chemical, and particulate inputs to the nearshore. Altered streamflows, which arrive at the shore at discrete locations (river mouths), from non-point source runoff, and from groundwater, are dispersed into bays, estuaries, and open coasts by nearshore hydrodynamic processes. Human modifications of the shore include armoring of the shoreline, engineered structures, filling or diking of marshes, and removal of riparian vegetation. These disturbances are diffuse spatially, but have direct effects at the land-sea interface, particularly on the physical energy environment, that may alter beach and nearshore processes over large spatial scales. The cumulative effects of physical and chemical changes ultimately are expressed as changes in nearshore habitat characteristics, biotic assemblages, and ecological processes.

Conceptual models like the one illustrated in figure 1-3 show how ecosystem components and processes are organized, how they interact, and how they respond to change. The models are working hypotheses about the structure and function of ecosystems because they are based on assumptions (Manley and others, 2000), and they help to identify gaps in knowledge, articulate hypotheses, and to test the strength of linkages.

[1] U.S. Geological Survey, Pacific Coastal and Marine Science Center, 400 Natural Bridges Drive, Santa Cruz, CA 95060.

[2] U.S. Geological Survey, Washington Water Science Center, 934 Broadway, Tacoma, WA 98402.

[3] U.S. Geological Survey, Western Fisheries Research Center, Columbia River Research Laboratory, 5501-A Cook-Underwood Road, Cook, WA 98605.

[4] U.S. Geological Survey, Rocky Mountain Geographic Science Center, 2150 Centre Avenue, MS 516, Fort Collins, CO 80526 (retired).

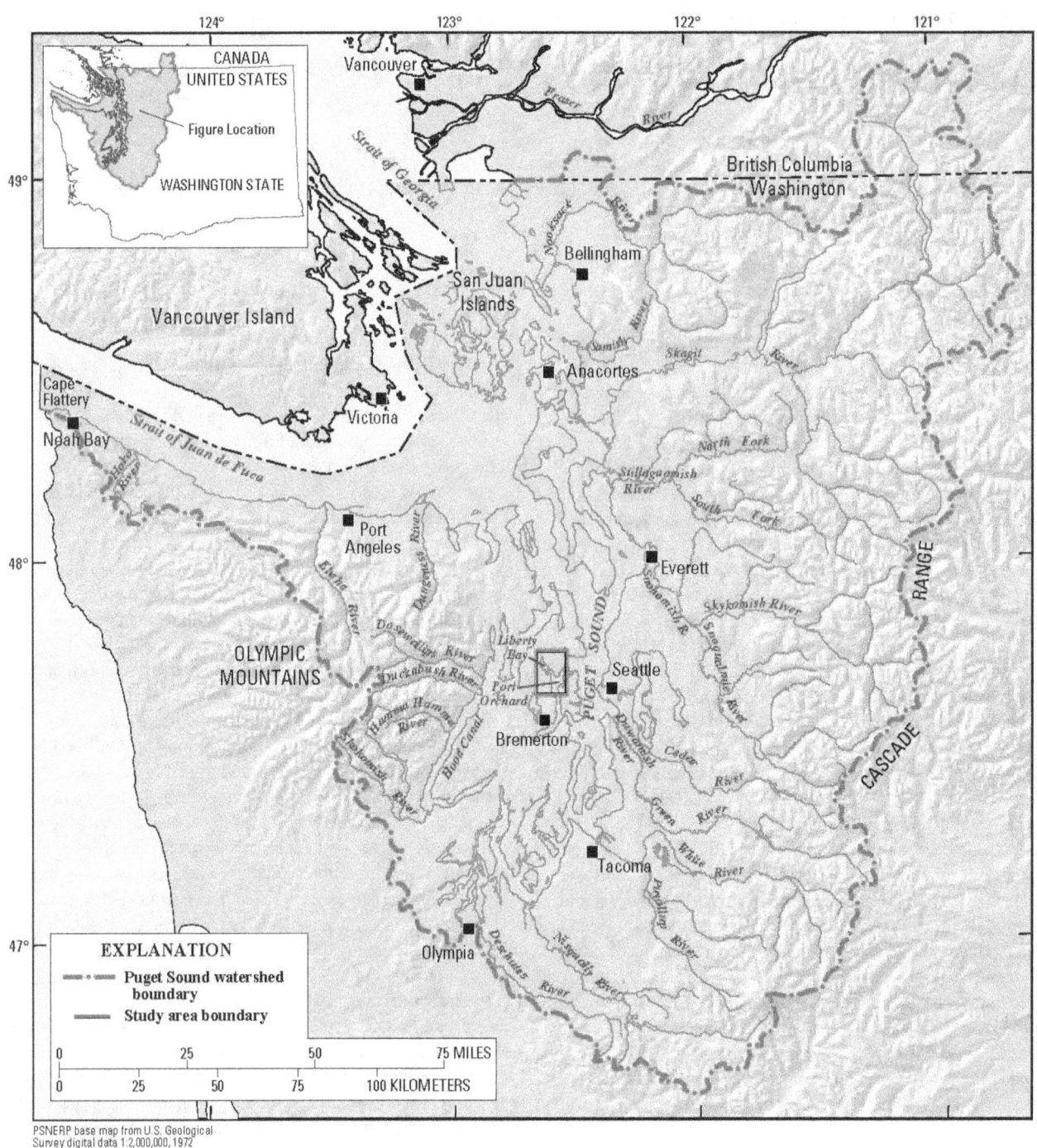

PSNERP base map from U.S. Geological
Survey digital data 1:2,000,000, 1972
Albers Equal-Area Conic Projection
Standard parallels 47° and 49°, central meridian 122°
Washington shaded relief, USGS, 30 meter DEM
British Columbia shaded relief, NASA, SRTM 90 meter

Figure 1-1.   Puget Sound estuary and drainage basin and the location of the study site in Liberty Bay, Kitsap County, Washington. Modified from Gelfenbaum and others (2006).

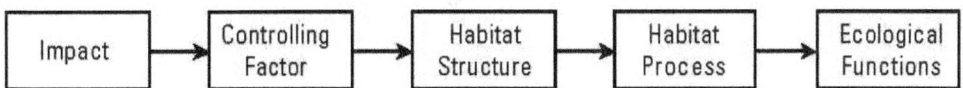

Figure 1-2. Conceptual model of mechanisms by which an unnatural disturbance, or impact, may alter ecosystem functions. From Williams and Thom (2001).

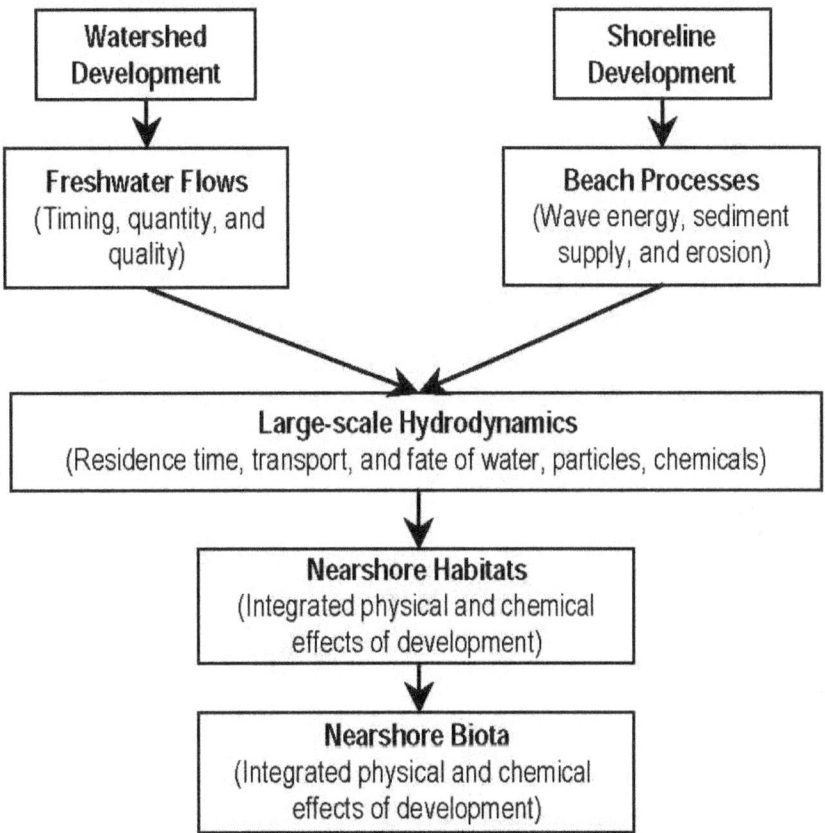

Figure 1-3. Conceptual system model of two pathways—watershed and shoreline development—through which human activities may affect nearshore ecosystems.

Purpose and Scope

In 2006 (April–May) and 2007 (January and May), a small-scale proof-of-concept (pilot) study investigated how contaminant inputs from an urban watershed and physical modifications of the shoreline affected conditions and processes of nearshore habitats. The first objective was to gain an understanding of the study site by describing nearshore physical, chemical, and biological attributes and processes. The second objective was to identify inputs of urban chemicals to the nearshore. The third objective was to identify nearshore ecological effects associated with urban development, empirically and with a predictive model.

Approach

To maximize the likelihood of detecting inputs from urban watersheds, the pilot study was conducted in an embayment. Dilution of riverine inputs with marine water was assumed to be less in an embayment than at the open shore, and water resident times, and consequently exposure times of biota, were assumed to be greater in an embayment than at the open shore. To identify altered habitats and processes associated with urban development, chemical and biological characteristics of an urbanized site were compared to those at a nearby site that was relatively non-urbanized. Quantitative differences between modified and unmodified sites were interpreted as possible consequences of urban watershed development (Goodsell and others, 2007).

Urban development in coastal watersheds and along shorelines may affect nearshore ecosystems in many different ways. Forage fish and their habitat needs were used as one measure of ecological health. Forage fish are the basis of Puget Sound food webs that include predatory fish, sea birds, and marine mammals (fig. 1-4). The life cycles of forage fish are closely tied to beach and nearshore habitats. Herring, for example, spawn on subtidal vegetation, preferably eelgrass, or other submerged objects. The eggs may be directly exposed to urban contaminants in water and sediment (fig. 1-4), which may cause developmental abnormalities or may increase susceptibility to parasites and disease. Juvenile and adult forage fish also may be exposed to contaminants in water and sediment if they are resident in an urbanized site. Obligate surf-spawning forage fish such as surf smelt and sand lance may be affected by physical modification of the shoreline because they require sand or gravel near high-tide elevations

to lay eggs. Overhanging back-beach vegetation and groundwater seeping out of the beach face protect eggs from desiccation and overheating (Rice, 2006). Because forage fish depend on several aspects of the beach and nearshore environment, forage fish and their habitat needs are a sensitive indicator of the effects of urbanization on the nearshore ecosystem.

Study Site

Liberty Bay, in Kitsap County, Washington, was selected as the pilot study site based on several criteria: (1) a gradient in urban development, (2) a broad mix of coastal land-uses and degrees of shoreline armoring, (3) a history of sewage spills and impaired water quality, (4) the proximity of herring spawning grounds and pre-spawn holding areas, and (5) documented sand lance and surf smelt spawning beaches. This combination of characteristics was found in few other embayments in Puget Sound.

Liberty Bay is a 7-km long, 12-m deep embayment on the western shore of Kitsap County in Central Puget Sound, Washington (fig. 1-5). Liberty Bay exchanges nearshore water with Port Orchard and is prevented from direct exchange with the main channel of Puget Sound by Bainbridge Island, to the east. Keyport and Lemolo peninsulas, which converge to within 200 m of each other at the mouth of Liberty Bay, protect the bay from southerly wind-driven waves.

Development of Liberty Bay and its watershed began in the late 1880s with the logging industry. By the early 1900s, the forests were depleted, and commercial fishing, shellfish, and agriculture industries flourished. More than 80 ha of tidelands around Liberty Bay were used for oyster production. By 1967, however, water quality had deteriorated to the point that the oyster beds on the eastern shore of Liberty Bay were closed to harvesting. Oyster production ceased entirely with the closing of the shucking plant in 1983; in 1993 Coast Seafoods Company sold its tidelands to the Washington State Department of Fish and Wildlife. Industrial development in Liberty Bay was associated with operations at the U.S. Naval Undersea Warfare Center (NUWC), Division Keyport, established in 1914. Welding, metal plating, painting, fuel storage, sludge disposal, and landfilling may have contributed to elevated heavy metals, hydrocarbon residues, polychlorinated biphenyls (PCBs), and volatile organic compounds in soil, sediment, and wastewater (Pine, 1975). The NUWC is now a U.S. Environmental Protection Agency superfund site.

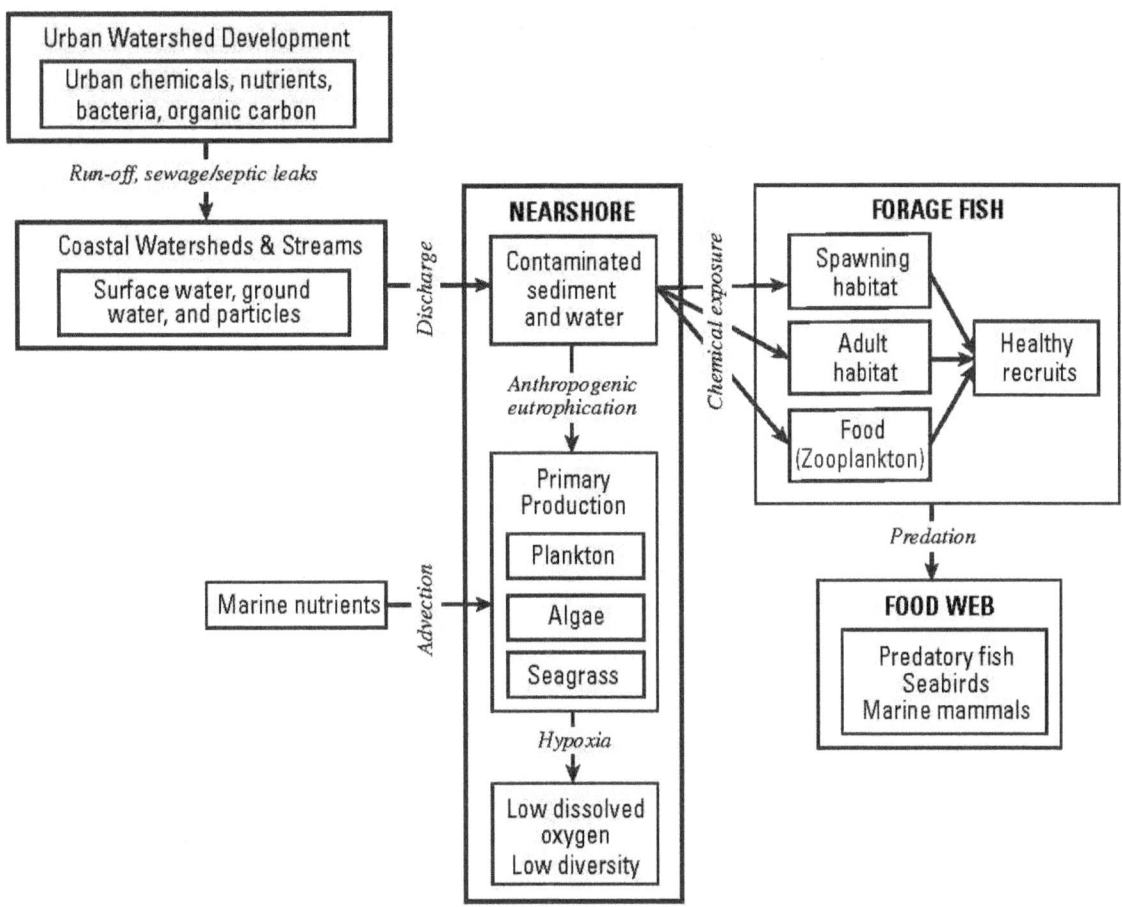

Figure 1-4. Conceptual model of urban-contaminant transport-pathways from sources in a coastal watershed, contaminant transfer through the food chain, and contaminant accumulation in forage fish and other nearshore organisms.

Eleven creeks flow into Liberty Bay. The Kitsap County Health District monitors water quality parameters for six of these creeks annually; none of the creeks meets Washington State water quality standards (Kitsap County Health District, 2006). Dogfish Creek is the largest stream flowing into the bay (mean annual discharge is 0.3 m³/s and peak discharge is 4.8 m³/s between October 1996 and September 1998) and its watershed includes the main urban center of Poulsbo (population 7,490). About one-half of the watershed is forested

(May and others, 2005). The remainder consists of urban and rural residential neighborhoods, commercial and light industrial centers, agricultural land, tribal land, and the naval base on Keyport peninsula (May and others, 2005). Like most areas around Puget Sound, major commercial and residential growth is occurring in the city of Poulsbo. The results of this study may help decision-makers balance the natural resource needs of human populations and the nearshore ecosystem.

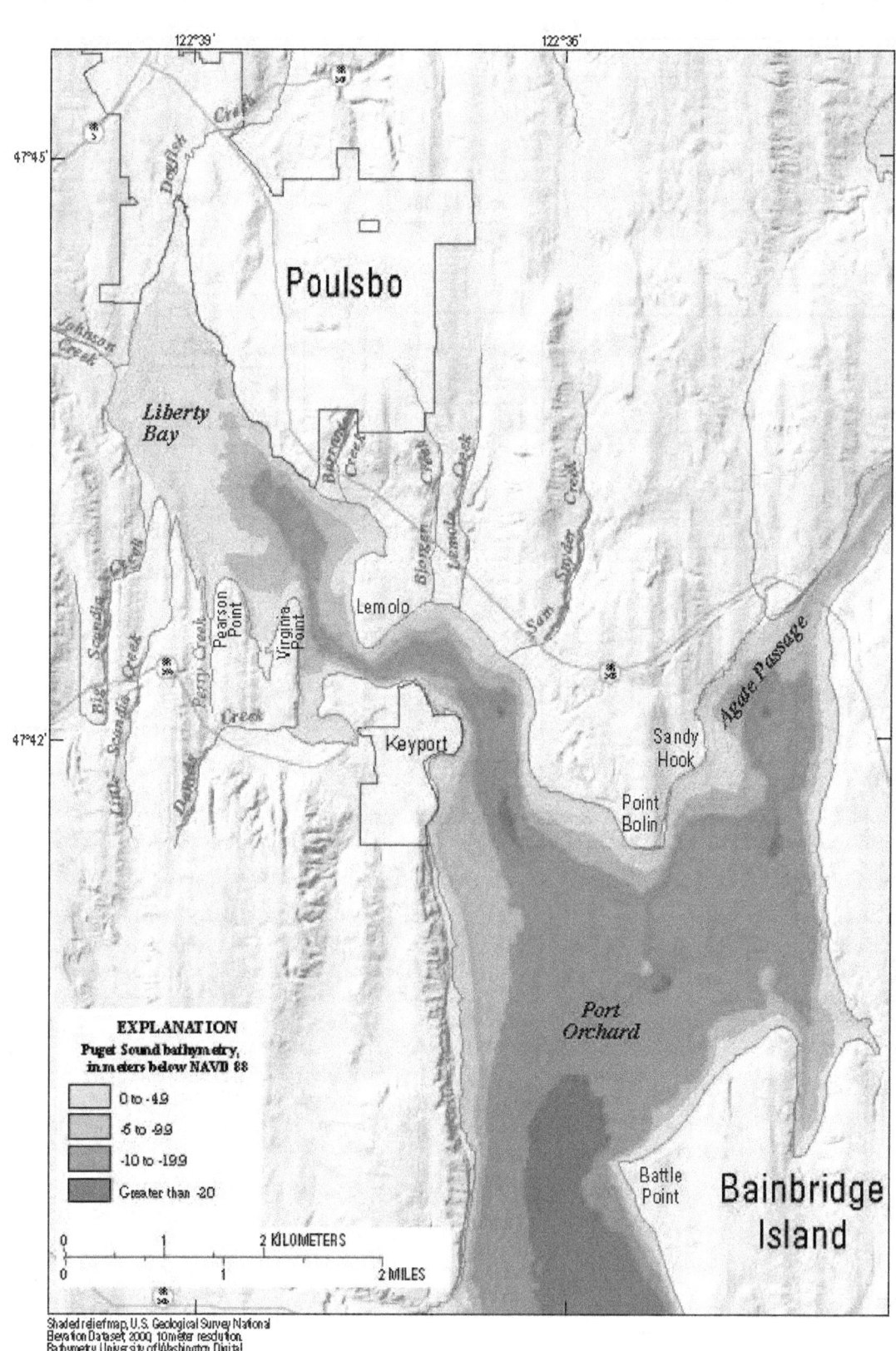

Figure 1-5. Liberty Bay and vicinity, Kitsap County, Washington.

References Cited

Gelfenbaum, Guy, Mumford, Tom, Brennan, Jim, Case, Harvey, Dethier, Megan, Fresh, Kurt, Goetz, Fred, van Heeswijk, Marijke, Leschine, Thomas M., Logsdon, Miles, Myers, Doug, Newton, Jan, Shipman, Hugh, Simenstad, C.A., Tanner, Curtis, and Woodson, David, 2006, Coastal Habitats in Puget Sound—A research plan in support of the Puget Sound Nearshore Partnership: Puget Sound Nearshore Partnership Report 2006-1, U.S. Geological Survey 2006-1, 46 p.

Goodsell, P.J., Chapman, M.G., and Underwood, A.J., 2007, Differences between biota in anthropogenically fragmented habitats and in naturally patchy habitats: Marine Ecology Progress Series, v. 351, p. 15–23.

Kitsap County Health District, 2006, Liberty Bay/Miller Bay watershed 2006 water quality monitoring report: Bremerton, Wash., Kitsap County Health District, 14 p.

MacDonald, Keith, Simpson, David, Paulson, Bradley, Cox, Jack, and Gendron, Jane, 1994, Shoreline armoring effects on physical coastal processes in Puget Sound, Washington: Olympia, Wash., Washington Department of Ecology, Shorelands and Water Resources Program, 171 p.

Manley, P.N., Zielinski, W.J., Stuart, C.M., Keane, J.J., Lind, A.J., Brown, Cathy, Plymale, B.L., and Napper, C.O., 2000, Monitoring ecosystems in the Sierra Nevada—The conceptual model foundation: Environmental Monitoring and Assessment, v. 64, p. 139–152.

May, C.W., Byrne-Barrantes, Kathleen, and Barrantes, L.E., 2005, Liberty Bay nearshore habitat evaluation and enhancement project final report: Poulsbo, Wash., Prepared for the Lemolo Citizens Club and Liberty Bay Foundation, p. 168.

Pine, Ron, 1975, Liberty Bay heavy metal problem—Status report, memo to Glen Fiedler and John Spencer: Olympia, Wash., Washington State Department of Ecology, Publication 75-e50, p. 39.

Rice, C.A., 2006, Effects of shoreline modification on a northern Puget Sound beach—Microclimate and embryo mortality in surf smelt (Hypomesus pretiosus): Estuaries and Coasts, v. 29, no. 1, p. 63–71.

Thom, R.M., Shreffer, D.K., and MacDonald, Keith, 1994, Shoreline armoring effects on coastal ecology and biological resources in Puget Sound, Washington: Olympia, Wash., Washington Department of Ecology, Shorelands and Water Resources Program, 102 p.

Williams, G.D., and Thom, R.M., 2001, Marine and estuarine shoreline modification issues: Richland, Wash., Pacific Northwest National Laboratory, Battelle Marine Sciences Laboratory, 136 p.

Suggested Citation

Takesue, R.K., Dinicola, R.S., Lacy, J.R., Liedtke, T.L., Rondorf, D.W., Smith, C.D., and Watts, R.D., 2011, Overview of effects of urbanization on the nearshore ecosystem of Puget Sound, Washington, chap. 1 of Takesue, R.K., ed., Hydrography of and biogeochemical inputs to Liberty Bay, a small urban embayment in Puget Sound, Washington: U.S. Geological Survey Scientific Investigations Report 2011–5152, p. 1–8.

Chapter 2. Aquatic Environment: Circulation, Water Quality, and Phytoplankton Concentration

By Jessica R. Lacy[1] and Richard S. Dinicola[2]

Introduction

Hydrodynamic and water quality measurements were made in and near Liberty Bay (fig. 2-1) to determine whether the properties or quality of waters within Liberty Bay differ from those of waters outside Liberty Bay in Port Orchard, and to evaluate whether any detected differences are related to anthropogenic inputs or modifications. Measurements focused on properties related to turbidity, eutrophication or eutrophication potential, and tidal mixing. The effect of anthropogenic inputs on receiving waters depends both on the mass or volume input and the residence time (or conversely dilution) of the receiving waters. Spatial and tidal-cycle variability in currents were measured in the study area on May 1, 2006, to assess circulation patterns and residence time in the study area. During April and May 2006, marine surface-layer temperature, salinity, suspended solids concentration, and fluorescence (an indicator of phytoplankton concentration) were measured continuously at moorings in the middle of Liberty Bay (station LB) and outside Liberty Bay at Point Bolin (station PB). Vertical profiles of these same properties were measured along the axis of Liberty Bay out to Point Bolin on one day during the deployment to better characterize their spatial variability. Nutrient (phosphate, silicate, nitrate, nitrite, and ammonia) concentrations and other water quality characteristics (temperature, dissolved oxygen, pH, and specific conductance) were measured weekly in surface and bottom waters at LB and PB in April and May 2006.

Current Measurements

Current speed and direction were measured with a boat-mounted acoustic Doppler current profiler (ADCP) in seven transects across Liberty Bay or Port Orchard during spring tides on May 1, 2006. The 1,200-kHz ADCP measured vertical velocity profiles with 50-cm depth resolution. The result, for each transect, is a snapshot of the vertical and cross-transect distribution of current speed and direction. The data were collected to evaluate the magnitude of tidal currents, and general circulation patterns in the study area. The transects were measured during spring tides, when tidal currents are strongest, to evaluate the potential for tidal mixing and flushing in enclosed embayments such as Liberty Bay. The location, timing, and tidal phase of the transects are shown in figure 2-2. Most of the data were collected during a strong flood tide. Data collected along the transect adjacent to Keyport at three points in the tidal cycle illustrate variation over the tidal cycle. Results of all ADCP transects are shown in appendix A figures.

The two transects measured at the end of the ebb tide showed that southward currents slowed from 0.6 m/s to approximately 0.1 m/s in the distance between the Keyport and Point Bolin transects, due to channel widening. On the western side of the channel adjacent to Keyport weak flow was directed northwards into Liberty Bay, although the tide was still ebbing on the eastern side of the transect (fig. A2-3).

Depth-averaged currents during flood tide along four transects (fig. 2-3) show some features of circulation influencing Liberty Bay. Flow through Agate Pass into Port Orchard is focused in the western half of the deep channel, where depth-averaged currents exceed 1 m/s. Tidal currents east of Battle Point, where Port Orchard is deep and wide, were much weaker (about 0.1 m/s), and directed to the north. Thus, for at least part of the flood tide, flows from Agate Passage and southern Port Orchard converge near Point Bolin, and water from both sources flows into Liberty Bay. Flood-tide current speed was about 0.35 m/s at Keyport and about 0.65 m/s in the channel near station LB (see fig. 2-1 for locations). These results indicate that spring-tide tidal excursion in the channel of Liberty Bay is approximately 7 km, assuming a spatially averaged maximum tidal current of 0.5 m/s, sinusoidal tidal currents, and a 12.48 hour tidal period. With a 7-km tidal excursion, passive particles starting at Poulsbo are transported out of Liberty Bay during each strong ebb tide.

[1]U.S. Geological Survey, Pacific Coastal and Marine Science Center, 400 Natural Bridges Drive, Santa Cruz, CA 95060.

[2]U.S. Geological Survey, Washington Water Science Center, 934 Broadway, Tacoma, WA 98402.

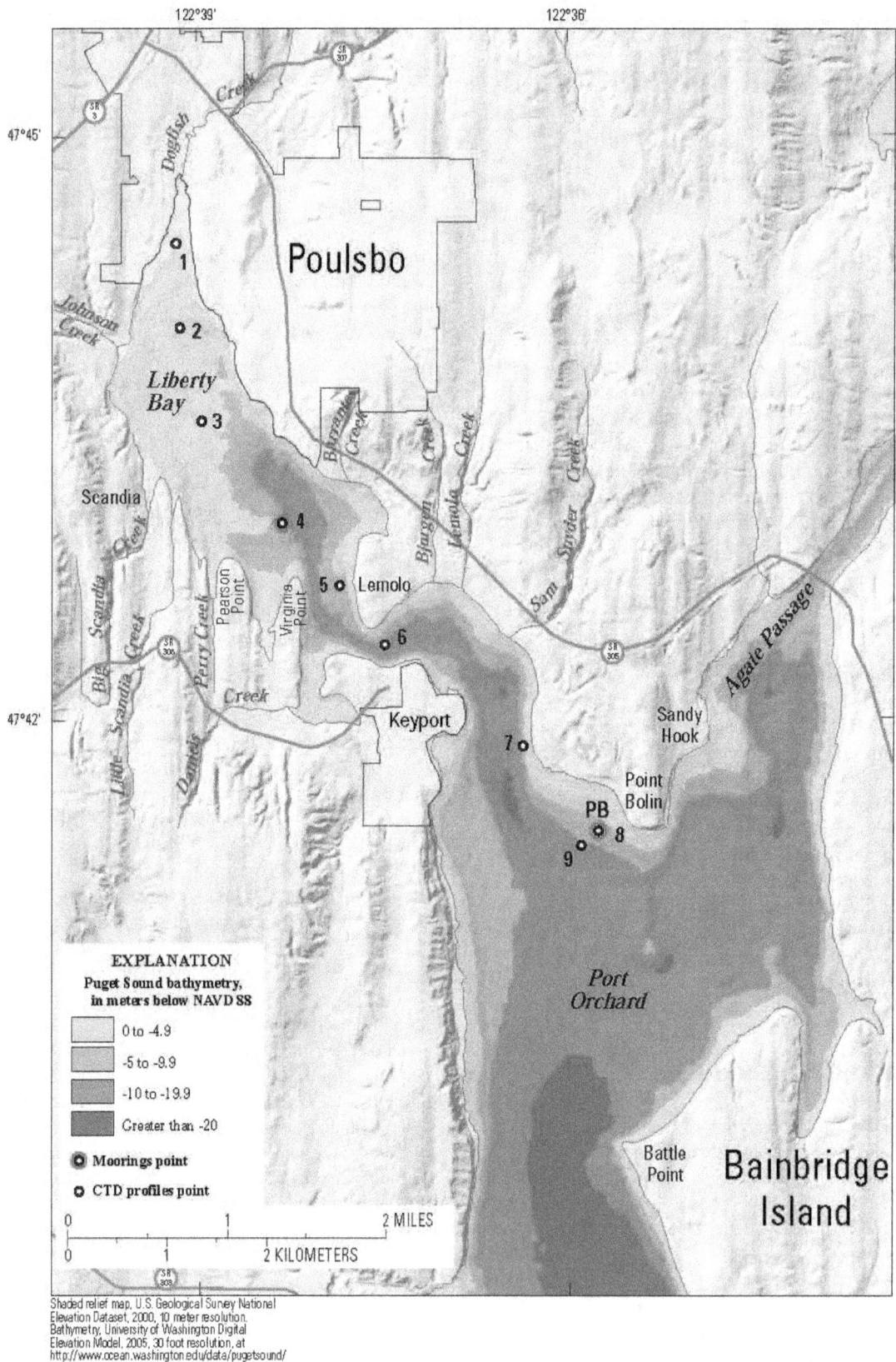

Figure 2-1. Locations of surface buoys for time-series data collection, conductivity, temperature, and depth sensor (CTD) profiling stations, Liberty Bay and vicinity, Kitsap County, Washington.

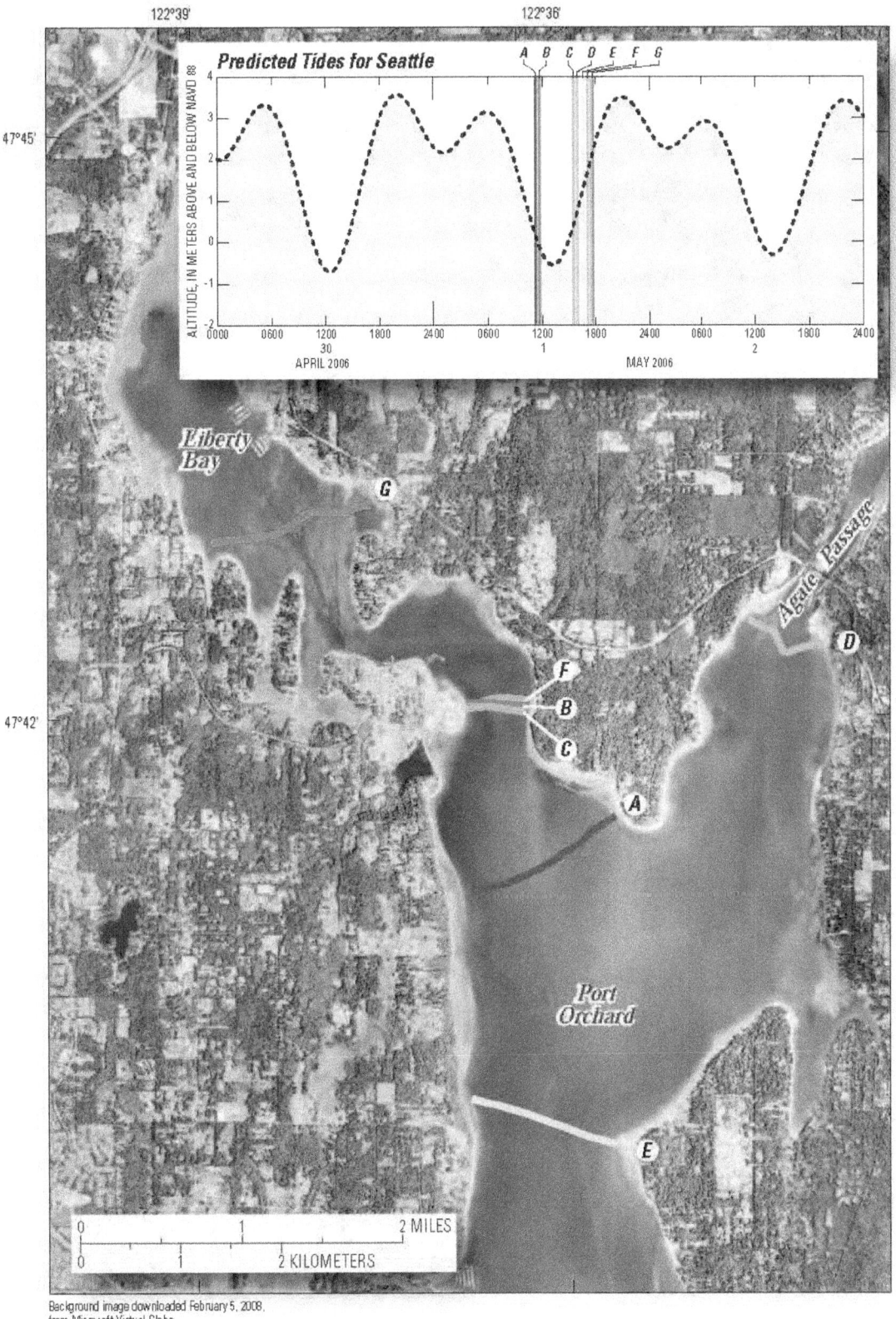

Figure 2-2.   Locations and tidal phases of acoustic Doppler current profiler (ADCP) transects on May 1, 2006, Liberty Bay and vicinity, Kitsap County, Washington.

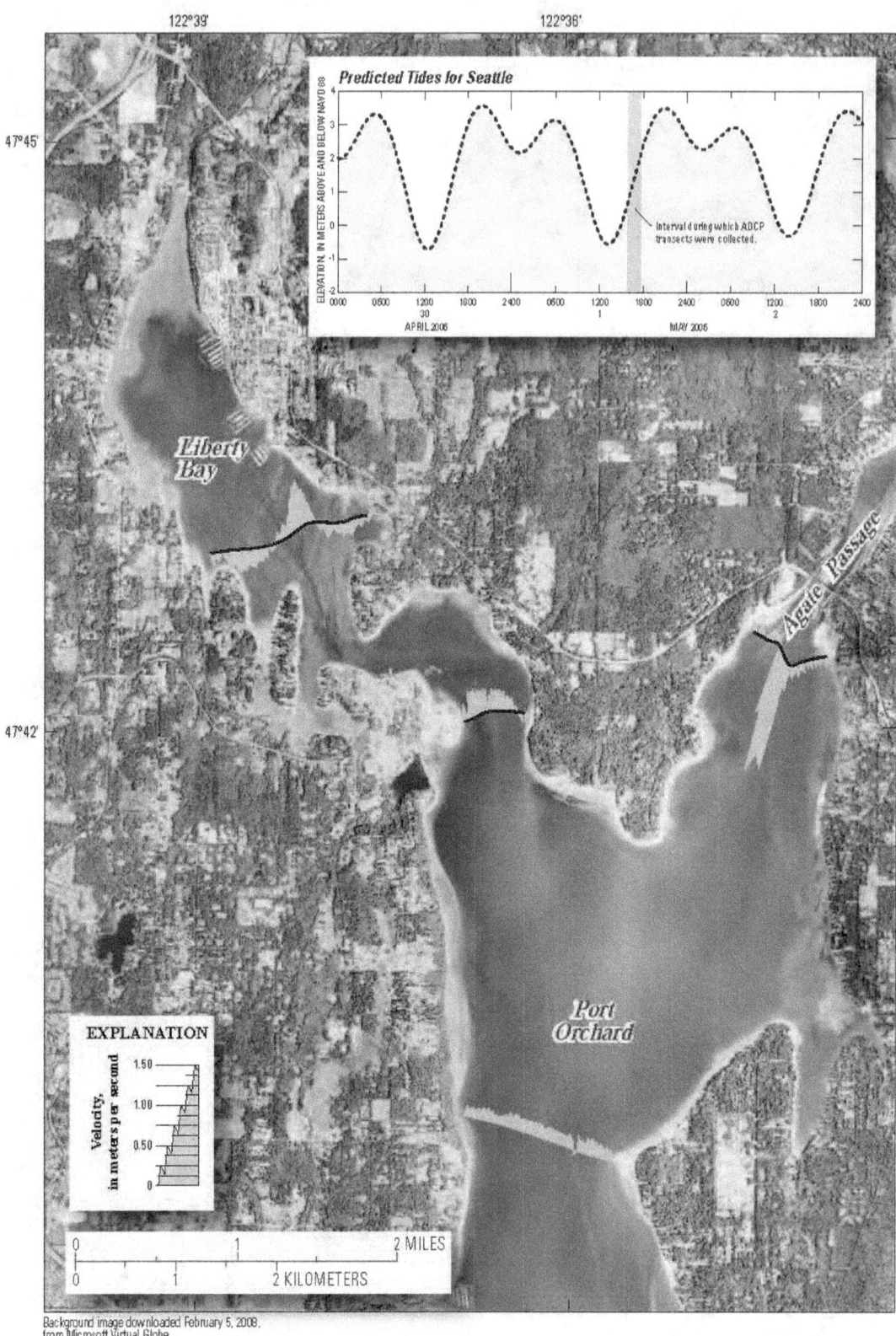

Background image downloaded February 5, 2008,
from Microsoft Virtual Globe.

Figure 2-3.   Depth-averaged currents from four acoustic Doppler current profiler (ADCP) transects during flood tide on April 1, 2006, Liberty Bay and vicinity, Kitsap County, Washington.

Currents are slower and residence times are longer in the shallows of Liberty Bay. The ADCP transect adjacent to the LB station shows that recirculation eddies form in the small indentations along the shore of Liberty Bay, which could retain water and particles, including sediment, detritus, phytoplankton, and zooplankton. Tidal currents were not measured during neap tides, but tidal excursions for neap tides likely are about one-half the tidal excursions of spring tides, which would result in much longer residence times in Liberty Bay during neap than spring tides.

In-situ Time Series and Calibration Data

Temperature, salinity, suspended-solids concentration (SSC), and fluorescence were measured at 10-minute intervals at LB and PB during April and May (fig. 2-1). The data were collected with SEACAT 16+ or SEACAT 16 (Sea-Bird Electronics) conductivity and temperature sensors (CT) at stations LB and PB, respectively. The CTs also logged two external sensors: an optical backscatter sensor (OBS, D&A Instruments) and a fluorometer (Cyclops model, chlorophyll *a* fluorescence, Turner Designs). Optical backscatterance is linearly related to SSC (Downing and others, 1981). Fluorescence commonly is used as an indicator of chlorophyll *a* concentration, and thus phytoplankton standing stock. The term indicator is used because the relationships between fluorescence, chlorophyll *a* concentration, and phytoplankton biomass vary with phytoplankton species, condition, and environmental factors (Kiefer, 1973). Nevertheless, *in-situ* fluorescence has been used widely and successfully to determine spatial distribution and magnitude of phytoplankton standing stocks (Welschmeyer, 1994).

The CTs were deployed from surface moorings, with all sensors approximately 1 m below the surface, from March 31 to May 31, 2006. The OBS and fluorometers were cleaned at 7 to 10 day intervals; nevertheless, some of the data were degraded by biofouling. Spikes, outliers, and data compromised by biofouling were removed from the time series. Additionally, there are no OBS data from LB for April 3–16 because of instrument malfunction.

Each time the *in-situ* sensors were cleaned, discrete water samples were collected to analyze for total suspended solids (TSS) and chlorophyll *a* to calibrate the OBS and fluorometers, respectively. Samples were collected 1 m below the surface with a Van Dorn sampler. Samples for chlorophyll *a* analysis were stored in dark bottles, and filtered within 2 hours of collection. Filters were stored at -80 °C (or on dry ice when transported) until analyzed. Filters were digested in 90 percent acetone overnight and centrifuged, and the solute was analyzed with a Turner TD700 fluorometer. TSS samples were filtered within 24-hours of collection, and filters were dried and weighed to determine solids concentration. Calibration equations for converting measured voltages to either chlorophyll *a* or SSC were determined by linear regression between concentrations measured in the discrete samples and voltages synchronously measured by the *in-situ* sensors (fig. 2-4).

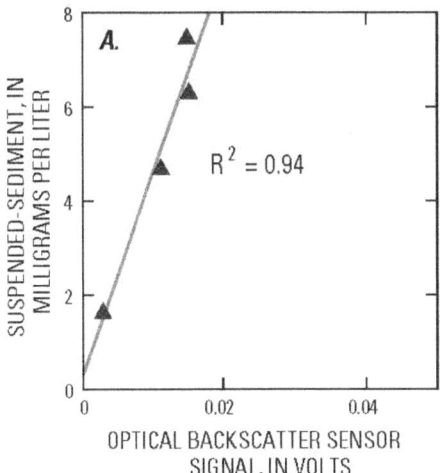

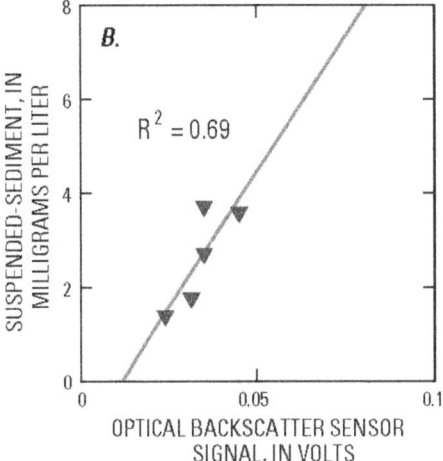

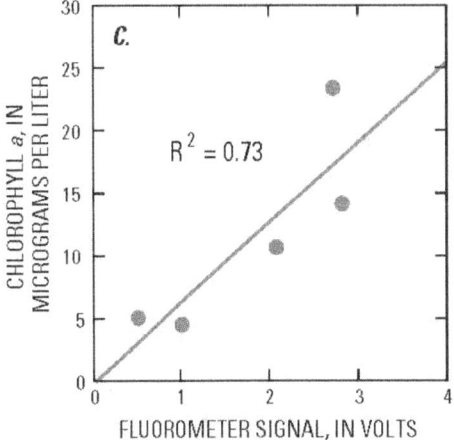

Figure 2-4. Calibration curves for suspended-solids concentration at (*A*) LB, (*B*) PB, and (*C*) chlorophyll *a* at LB, Kitsap County, Washington.

Surface temperatures increased from 10 to 14 °C during the deployment, and were 1-2 °C higher at LB than at PB. Salinities were between 28 and 29 practical salinity units (psu) with little spatial or temporal variability, indicating that freshwater inflows to Liberty Bay did not have a widespread effect on salinity during the study period (fig. 2-5).

SSC was greater at LB than at PB, and the range of SSC also was much greater at LB (fig. 2-6). SSC at LB varied with the tidal cycle, with pulses of high SSC during flooding tides after the lower-low tides. During these pulses, SSC was two to three times greater at LB than at PB. In contrast, there is no clear tidal signal in the SSC data from PB. At both sites, SSC increased at the same time that lower salinity water appeared at PB, on April 21, 2006.

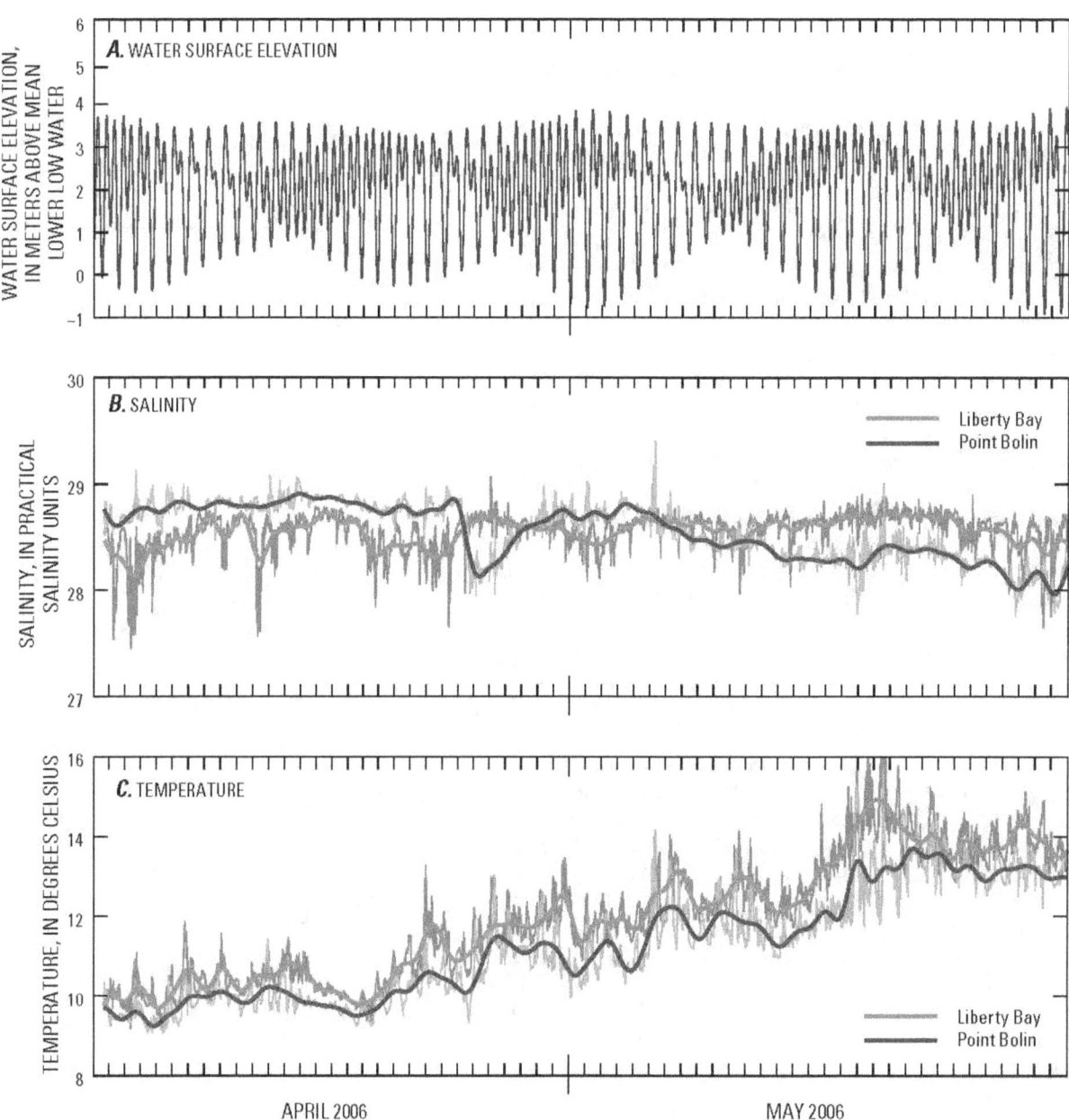

Figure 2-5.   Time series of (*A*) predicted tidal elevation at Seattle, (*B*) salinity, and (*C*) temperature at LB and at PB, Kitsap County, Washington. Thin lines are raw data with 10-minute sampling interval; thick lines are tidal averages.

The fluorometer voltages showed several peaks in chlorophyll *a*, indicating high concentrations of phytoplankton (blooms) at both sites during the 2-month deployment (fig. 2-7). The data also exhibited a strong diurnal signal, with lower fluorescence during daylight hours (fig. 2-7*B*). This daily fluctuation is attributed to photo-inhibition (or photo-quenching) of fluorescence, which occurs in response to strong irradiation in clear water near the surface (Abbott and others, 1982; Cullen and Lewis, 1995), and is thus an artifact of using fluorescence to measure phytoplankton concentration. In the plot of calibrated chlorophyll *a* concentration (fig. 2-7*C*), all measurements taken between 5 a.m. and 7 p.m. (local time), and data compromised by instrument fouling have been removed from the record.

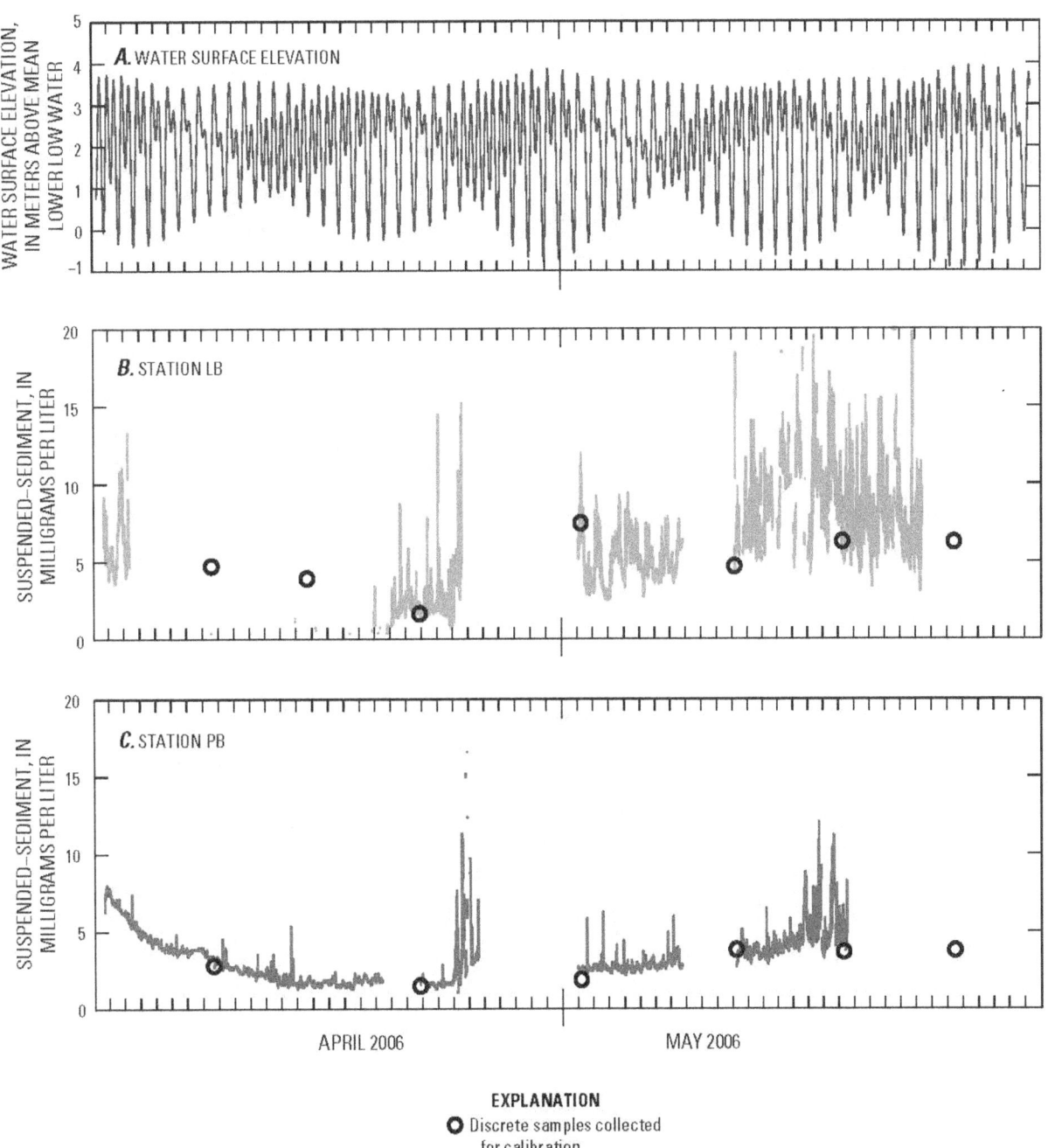

EXPLANATION

⬤ Discrete samples collected for calibration

Figure 2-6. Time series of (*A*) water-surface elevation, and suspended-solids concentration measured at (*B*) LB, and (*C*) PB, Kitsap County, Washington. Data compromised by instrument fouling have been removed.

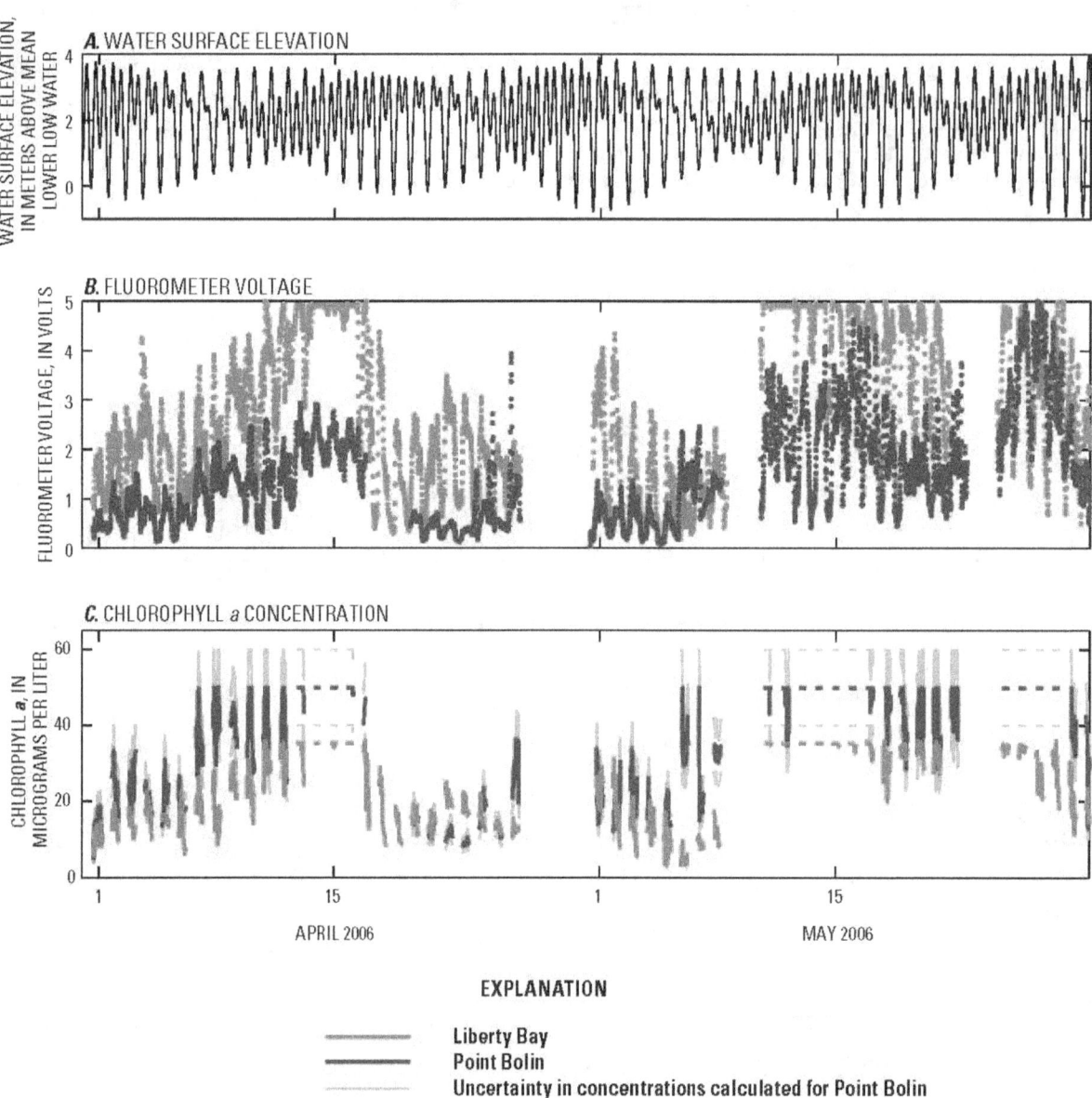

Figure 2-7.   Time series of (*A*) water-surface elevation, (*B*) fluorometer voltage, and (*C*) chlorophyll *a* concentration, Liberty Bay and Point Bolin, Kitsap County, Washington. Data compromised by fouling have been removed. Maximum measurable concentration (saturation value) was 36 µg/L at Liberty Bay and 50 µg/L at Point Bolin.

For station LB, a calibration equation for chlorophyll *a* concentration was derived from discrete samples using linear regression (fig. 2-4 *C*). At station PB, only four samples could be used for calibration, because of problems with instrument fouling and sample storage and analysis, and the coefficient of determination (R^2) for the regression relation was very low. To verify this relationship, we compared fluorescence voltages measured with the profiling conductivity, temperature, and depth sensor (CTD) (see Vertical Profiles section in this chapter) to voltages measured by the *in-situ* sensors at the two sites, and, in combination with the calibration data, determined a range of possible slopes between voltage and chlorophyll *a* concentration. The mean of the range was used for calculating chlorophyll *a* concentration. Because this method is more qualitative than the calibration for LB, upper and lower limits on chlorophyll *a* concentration at PB based on the range of possible slopes are shown in grey in figure 2-7 *C*. The 36 µg/L concentrations for LB and the 50 µg/L concentrations for PB in figure 2-7 *C* are maximum measurable concentrations; actual concentrations exceed these values by an unknown amount. At LB, this occurs when fluorescence voltages exceeded the saturation value of 5V, and at PB when fluorescence voltages were outside the range for which the improvised calibration method was valid. Although the uncalibrated voltages consistently were greater at LB than at PB, (fig. 2-7 *B*), this is not the case for the calibrated data (fig. 2-7 *C*) because of differences in gains for the two instruments.

Maximum chlorophyll *a* concentrations, indicating phytoplankton blooms, occurred around April 15 and May 15 at both stations. Bloom concentrations were at least five times greater than minimum measured concentrations. Chlorophyll *a* concentrations were greater at PB than at LB during blooms, whereas they were similar at the two sites when concentrations were low.

Vertical Profiles

To complement the time-series data, we collected water column profiles of temperature, salinity, turbidity, and chlorophyll *a* along the axis of Liberty Bay out to Point Bolin (see fig. 2-1 for profile locations) following high tide and low tide on April 30, 2006. The profiles provide additional information about horizontal and vertical variability of the monitored parameters. The data were collected with a SBE 19+ profiling CTD, with an OBS and a fluorometer as external sensors. Factory calibrations provided by SBE were used to convert optical backscatterance to turbidity (in nephelometric turbidity units or NTUs), and fluorescence to chlorophyll *a*. The factory calibrations do not incorporate site specific factors such as grain size and phytoplankton community that are taken into account in the field calibrations that were done for the time-series measurements. Additionally, turbidity is an optical property, which cannot be directly converted to SSC. For these reasons, the turbidity and chlorophyll *a* profiles are more useful for describing spatial variability than

for quantifying concentrations. Spatial and temporal variation in turbidity provides information about variation in SSC, because both are linearly related to OBS voltage, and in this system, SSC is considered the primary contributor to turbidity.

The profiles were collected during a period of spring tides (fig. 2-7 *A*), and between phytoplankton blooms (fig. 2-7 *C*). The profiling revealed a clear gradient in chlorophyll *a* concentration from the head of Liberty Bay to Point Bolin (fig. 2-8). Chlorophyll *a* was higher within Liberty Bay during the low-tide survey than during the high-tide survey, consistent with the decreased influence of waters (with lower chlorophyll *a* concentration) from outside the bay. Turbidity also was greater in Liberty Bay than at Point Bolin. At low tide, turbidity at the head of Liberty Bay was considerably higher than in any of the other profiles, and turbidity was elevated near the bottom in the inner part of Liberty Bay.

Four profiles taken at station LB illustrate the variation in the vertical structure over the tidal cycle (fig. 2-9). Turbidity is relatively low during the ebbing tide (08:10). Higher turbidity appears first at the bottom and then higher in the water column over the course of the flooding tide. The temporal pattern suggests that sediment is resuspended over the mudflats at low tide, and mixed upwards in the water column as tidal currents increase during the flood tide. In the chlorophyll *a* profiles, photo-quenching is evident in the 11:50 and 14:20 profiles, whereas concentration is fairly constant with depth in the first and last profiles.

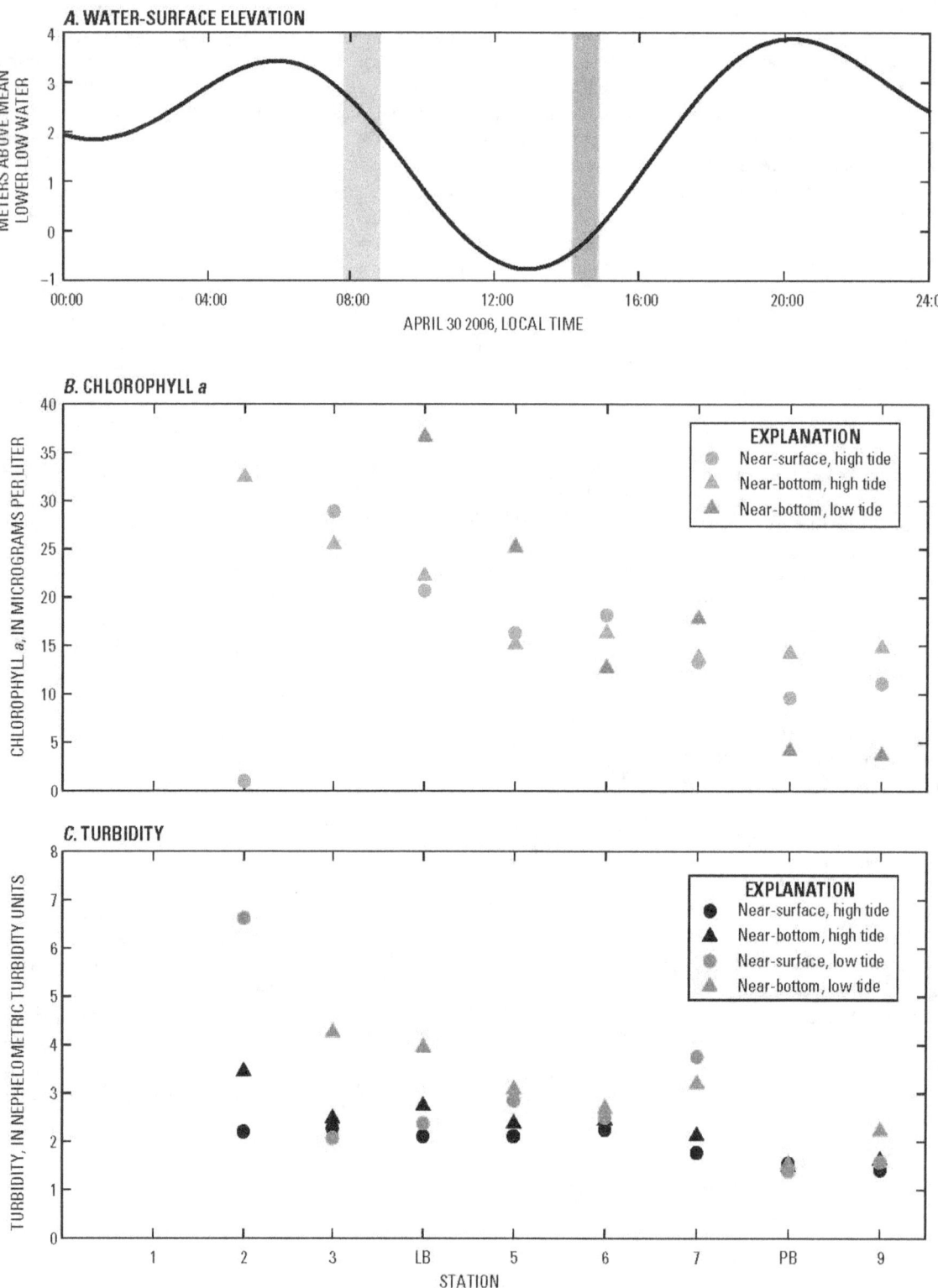

Figure 2-8.   Results from conductivity, temperature, and depth sensor (CTD) profiling on April 30, 2006, from Liberty Bay to Port Orchard, Kitsap County, Washington. (*A*) Predicted water-surface elevation showing interval of high tide (HT) (green), and low tide (LT) (magenta) profiles. Near-surface and near-bottom (*B*) chlorophyll *a* and (*C*) turbidity. Near-surface chlorophyll a concentrations during the LT survey were influenced by photo-quenching and are not shown.

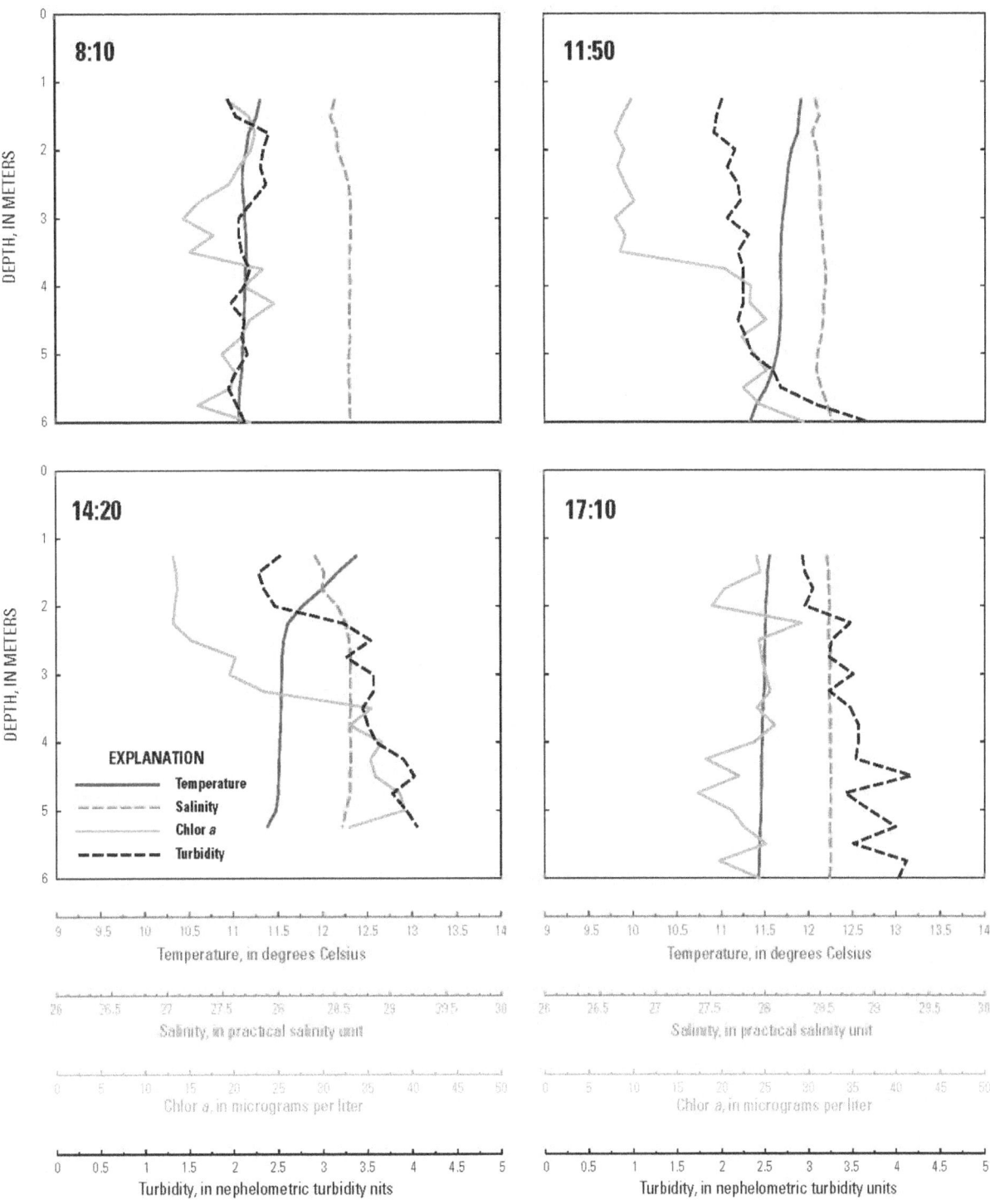

Figure 2-9. Profiles of temperature, salinity, chlorophyll *a* concentration, and turbidity at station LB, Kitsap County, Washington, April 30, 2006.

Nutrient Concentrations and Other Water-Quality Characteristics

Nutrient concentrations and other water-quality characteristics were measured in surface and bottom waters at LB and PB approximately weekly between April 7 and May 24, 2006. The goal was to characterize temporal variability in water quality, and to relate nutrient concentrations to observed chlorophyll *a* concentrations and inferred phytoplankton blooms. Samples were collected for analysis of the nutrients phosphate, silicate, nitrate, nitrite, and ammonia. Temperature, dissolved oxygen, pH, and specific conductance also were measured. Water samples were collected 1 m below the surface and 1 m above the seabed using a peristaltic pump and weighted (stainless steel) polyethylene tubing. Nutrient samples were filtered through 0.45-μm pore size surfactant-free cellulose filters into 125-mL brown polyethylene bottles, stored on ice, and shipped to the Marine Chemistry Lab at the University of Washington, School of Oceanography for analysis. Nutrients were analyzed on an Alpkem RFA/2 system following the protocols of the World Ocean Circulation Experiment Hydrographic Program

(2008). Dissolved oxygen (DO), water temperature, pH, and specific conductance were measured in the field in a flow-through chamber using a temperature-compensated YSI data sonde. On each sampling day, the DO sensor was calibrated using the water-saturated air method, the pH sensor was calibrated with two pH standards, and the specific conductance sensor was checked with standard reference solutions as described by Wilde (variously dated).

Nitrate concentrations in the surface layer at both sampling sites decreased steadily from about 150 to 6 μg/L during the study period (table 2-1 and fig. 2-10). Phosphate and silicate concentrations followed the same general trend (fig. 2-11), whereas ammonia concentrations did not (fig. 2-10). We attribute these temporal trends to enhanced phytoplankton productivity and nutrient uptake from spring into summer. Maximum chlorophyll *a* concentrations, indicating phytoplankton blooms, occurred around April 15 and May 15 at both stations (fig. 2-7). Nitrate concentrations in the bottom layer were greater than concentrations in the surface layer in all samples and did not decrease as consistently, suggesting that water column denitrification did not occur in spring or early summer.

Table 2-1. Nutrient concentrations and other marine water-quality characteristics in water column samples from sites at Liberty Bay and Point Bolin near Poulsbo, Washington, April and May 2006.

[Values in **bold** were less than concentrations reported for associated blank samples. **Abbreviations:** m, meters; MLLW, mean low-low water; mg/L, milligrams per liter; µS/cm, millisiemens per centimeter at 25 degrees Celsius; µg/L, micrograms per liter; —, not analyzed]

Date	Time	Sample depth	Tide (m above MLLW)	Dissolved phosphate (µg/L as P)	Dissolved silica (µg/L as Si)	Dissolved nitrate (µg/L as N)	Dissolved nitrite (µg/L as N)	Dissolved ammonia (µg/L as N)	Water temperature (degrees Celsius)	Dissolved oxygen (mg/L)	pH (units)	Specific conductance (µS/cm)	Total dissolved nitrogen (µg/L as N)
Liberty Bay surface layer													
04-07-06	11:06	1.0	2.1	38.1	700.3	152.3	4.0	6.1	10.7	11.4	7.65	4.400	162.3
04-13-06	12:05	1.0	0.4	31.3	846.3	57.7	2.1	9.7	10.2	10.2	7.85	4.423	69.5
04-20-06	13:05	1.0	1.5	36.5	1,126.2	44.6	2.0	26.9	12.2	9.8	8.00	4.380	73.6
05-10-06	11:40	1.0	0.6	22.2	0.0	**0.0**	**0.9**	**2.9**	13.8	14.0	8.00	4.584	3.7
05-17-06	11:33	1.0	0.9	3.1	55.1	4.4	0.1	0.7	15.3	13.6	8.19	4.171	5.2
05-24-06	11:43	1.0	0.5	14.7	86.6	5.8	0.3	4.3	14.5	12.7	7.76	4.108	10.4
Point Bolin surface layer													
04-07-06	12:55	1.0	2.4	39.3	866.2	130.6	4.1	10.2	11.1	11.7	7.84	4.465	144.8
04-13-06	—	—	—	—	—	—	—	—	—	—	—	—	0.0
04-20-06	11:45	1.0	2.1	36.8	1,183.4	81.9	**2.5**	22.1	11.7	10.7	7.87	4.426	106.5
05-10-06	12:38	1.0	1.1	17.1	797.5	**6.9**	**1.7**	**0.1**	13.0	8.9	8.00	4.684	8.8
05-17-06	12:22	1.0	0.3	17.6	476.6	41.7	2.2	5.8	13.7	11.7	7.97	4.212	49.7
05-24-06	12:37	1.0	1.0	7.0	81.5	6.0	0.1	1.7	13.7	13.7	8.01	4.147	7.8
Liberty Bay bottom layer													
04-07-06	11:15	8.2	2.1	48.5	953.1	215.4	4.3	21.1	10.3	10.4	7.32	4.450	240.8
04-13-06	12:40	6.7	0.4	38.0	890.1	85.6	2.4	38.1	10.3	9.0	7.91	4.445	126.1
04-20-06	13:15	7.6	1.5	43.2	1,204.9	87.9	**2.6**	45.1	11.3	10.5	7.87	4.448	135.7
05-10-06	11:59	6.1	0.6	45.5	1,124.2	107.5	3.9	59.2	13.2	10.2	7.80	4.645	170.6
05-17-06	10:17	6.7	1.9	13.3	132.9	7.0	0.1	1.2	14.0	12.0	7.91	4.186	8.3
05-24-06	11:58	6.7	0.5	28.7	144.7	11.1	0.4	27.4	13.7	11.8	7.80	4.138	38.9
Point Bolin bottom layer													
04-07-06	13:05	8.2	2.4	47.1	1,052.4	210.0	4.1	6.1	10.4	10.4	7.73	4.475	220.2
04-13-06	—	—	—	—	—	—	—	—	—	—	—	—	
04-20-06	11:50	7.6	2.1	44.9	1,336.1	159.0	**3.3**	25.0	10.6	11.1	7.66	4.445	187.3
05-10-06	12:55	6.4	1.1	45.2	1,223.9	166.6	**3.2**	42.9	11.9	3.2	7.80	4.714	212.7
05-17-06	12:08	6.4	0.3	30.4	648.1	117.7	2.4	29.6	12.6	10.9	7.88	4.211	149.7
05-24-06	12:43	6.7	1.0	11.4	147.0	15.4	0.4	5.3	13.3	13.2	8.00	4.155	21.1

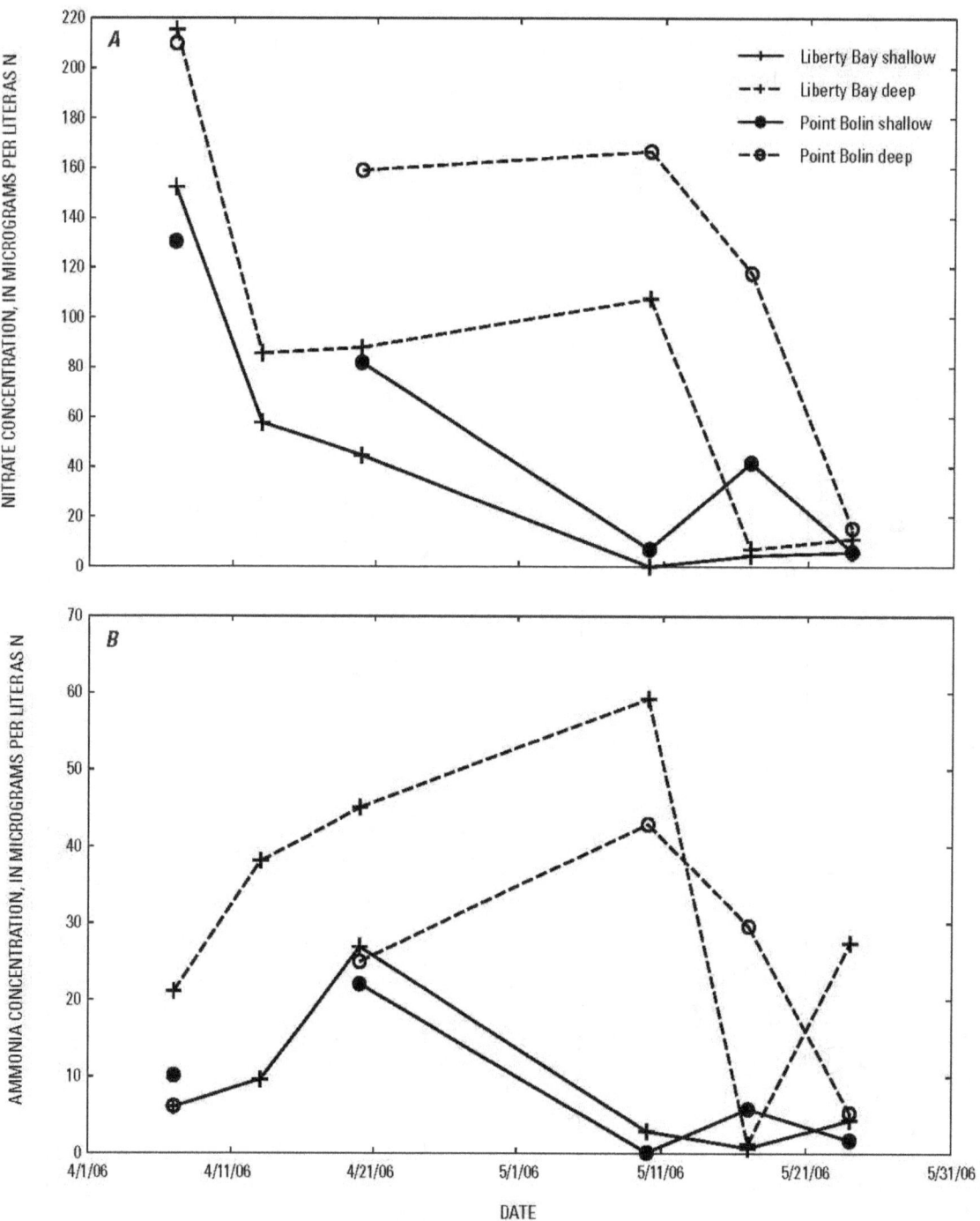

Figure 2-10.   (*A*) nitrate and (*B*) ammonia concentrations in weekly water samples from stations LB and PB, Kitsap County, Washington, April and May 2006.

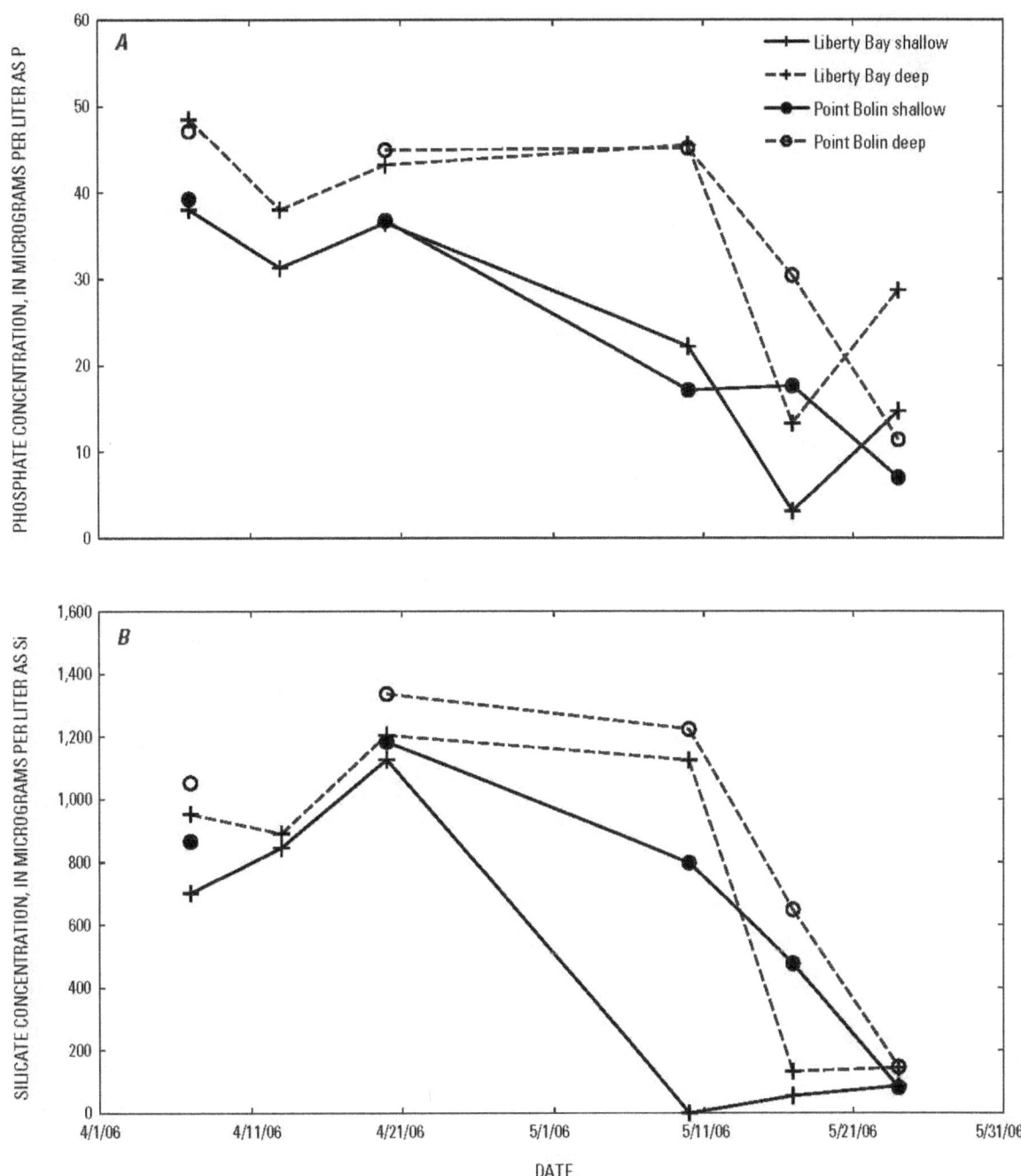

Figure 2-11. (A) phosphate and (B) silicate concentrations in weekly water samples from LB and PB, Kitsap County, Washington, April and May 2006.

Discussion and Conclusions

Surface waters at station LB inside Liberty Bay were more turbid and slightly warmer (0.5 to 1 °C) than at station PB near Point Bolin. Variability in turbidity due to SSC also was much greater at LB than at PB. The vertical profiles of turbidity taken on one day suggest that at depth, elevated turbidity in the interior of Liberty Bay (relative to PB) is even more pronounced, particularly at low tide. Both the SSC time-series and the turbidity profiles have peak values at the beginning of flood tides, suggesting that sediment is resuspended at low tide and retained in Liberty Bay by flood-tide currents. The turbidity levels in Liberty Bay may limit the distribution of eelgrass and other vascular plants as well as the vertical extent of phytoplankton blooms.

There were no consistent differences between the two stations in the other measured parameters: salinity, chlorophyll *a*, or nutrient concentrations. The lack of spatial gradients in these substances, as well as the current measurements, indicate that residence time in Liberty Bay is relatively short and that water is frequently exchanged between Liberty Bay and Port Orchard. The residence time estimated for most of Liberty Bay is on the order of one tidal cycle during spring tides, but is significantly longer during neap tides.

Phytoplankton blooms at the two stations were coincident. Data from both stations suggest that repeated phytoplankton blooms depleted dissolved nutrient concentrations during the 2-month study period. Nutrients were depleted earlier in the surface waters than at depth. Oxygen levels were close to saturation throughout the study period, with the exception of one measurement of 3.2 mg/L in the bottom waters at PB.

Nutrient and water-quality data from this study were compared to historical discrete data collected during April and May by the Puget Sound Ambient Monitoring Program (PSAMP) at two sites in Liberty Bay (State of Washington, variously dated). Near-surface and near-bottom data are available from PSAMP station POD006, near station LB, for 1991–92 and 1994–95, and from PSAMP station POD007, near the outlet of Dogfish Creek, for 1997–98 and 2004. For general water-quality parameters, the ranges in the data from LB and PB are very similar to those of the two PSAMP stations, with the exception of the very low DO concentration (3.2 mg/L) measured in the bottom water at station PB on May 10 (the lowest PSAMP April–May DO concentration was 7.5 mg/L). Nutrient concentrations in the two data sets are less consistent. The maximum nitrate and silicate concentrations measured in this study were substantially greater than those reported by PSAMP. In contrast, the maximum phosphate and ammonia concentrations measured at LB and PB were substantially less than those reported by PSAMP. PSAMP samples were collected approximately monthly, so the April–May period typically is represented by two data points; however, PSAMP has sampled multiple years. Comparison of the two data sets suggests that both short-term and interannual variability in marine water-quality can be high, and that sporadic synoptic measurements are insufficient for evaluating water quality or understanding nutrient loading or cycling.

PSAMP data throughout Puget Sound show an annual cycle in nutrient and chlorophyll *a* concentrations (Newton and others, 2002). In winter, nutrient concentrations are high, approaching marine concentrations, and chlorophyll *a* concentrations are low, because of limited light availability.

In spring and summer, chlorophyll *a* concentrations increase and nutrient concentrations decrease because of phytoplankton uptake. The April–May period of our study typically coincides with the transition from winter to summer conditions, which is consistent with the repeated blooms and declining nutrient concentrations in the data.

At open-water sites in central Puget Sound chlorophyll *a* concentrations in surface waters rarely exceed 15 µg/L (Mackas and Harrison, 1997). In monthly monitoring by the Washington Department of Ecology during 1998–2000, near-surface chlorophyll *a* concentrations at open-water sites, including PSAMP stations ADM001, ADM002, and PSB003, either never or only occasionally exceeded 15 µg/L (Newton and others, 2002). However, at enclosed sites with less exchange with marine waters, including Sinclair Inlet (PSAMP station SIN001), Oak Harbor (PSAMP station OAK004), and some locations in Hood Canal (PSAMP station HCB006), chlorophyll *a* concentrations between 20 and 40 µg/L were measured at least once a year (Newton and others, 2002). Thus, bloom concentrations observed at both the LB and PB stations are typical of the elevated levels of phytoplankton that occur in enclosed areas of Puget Sound. These elevated levels suggest that nutrient availability is greater at these sites; however, elevated phytoplankton concentrations also can be caused by physical factors such as decreased mixing of phytoplankton out of the photic zone (Newton and others, 2002). The time-series observations from Liberty Bay show that bloom duration typically was about a week, so that blooms would be captured as single points or missed entirely, by a monthly monitoring program such as PSAMP. The data also show that phytoplankton concentrations can vary widely from week to week, and that repeated blooms occurred over the period that nutrient concentrations decreased.

Profiles taken on a day between phytoplankton blooms showed a gradient in chlorophyll *a* concentration, with higher concentrations in Liberty Bay than at the mouth of the bay near Point Bolin. This suggests that although system-wide blooms are controlled by availability of marine nutrients, local anthropogenic nutrient sources in Liberty Bay may maintain a higher baseline of phytoplankton concentration in Liberty Bay than outside the bay in Port Orchard. Estimates of dissolved inorganic nutrient (DIN) flux in submerged groundwater discharge (see chapter 3 of this report) show greater DIN flux in Liberty Bay than at Sandy Hook on Agate Pass, and local creeks likely are additional sources of nutrient loading to Liberty Bay. A longer data set, extending further into the summer period of nutrient depletion, would help to assess the importance of local nutrient sources to phytoplankton dynamics.

Acknowledgments

Thanks to Greg Justin and Chris Curran of the U.S. Geological Survey, Luis Escalante and Luis Barrantes of the Liberty Bay Foundation, and Paul Dorn of the Suquamish Tribe for assistance with data collection, and to Andrew Stevens and Theresa Olsen of the U.S. Geological Survey for preparing figures. Tara Schraga of the USGS provided training and laboratory facilities for analysis of chlorophyll samples. This chapter was improved through reviews by Richard Wagner and Curt Storlazzi of the USGS.

References Cited

Abbott, M.R., Richerson, P.J., and Powell, T.M., 1982, In situ response of phytoplankton fluorescence to rapid variations in light: Limnology and Oceanography, v. 27, no. 2, p. 218-225.

Cullen, J.J., and Lewis, M.R., 1995, Biological processes and optical measurements near the sea surface—Some issues relevant to remote sensing: Journal of Geophysical Research, v. 100, no. C7, p. 13,255-13,266.

Downing, J.P., Sternberg, R.W., and Lister, C.R.B., 1981, New instrumentation for the investigation of sediment suspension processes in the shallow marine environment: Marine Geology, v. 42, p. 19-34.

Kiefer, D.A., 1973, Fluorescence properties of natural phytoplankton populations: Marine Biology, v. 22, p. 263-269.

Mackas, D.L., and Harrison, P.J., 1997, Nitrogenous nutrient sources and sinks in the Juan de Fuca Strait/Strait of Georgia/Puget Sound estuarine system—Assessing the potential for eutrophication: Estuarine, Coastal and Shelf Science, v. 44, no. 1, p. 1-21. (Also available at http://dx.doi.org/10.1006/ecss.1996.0110.)

Newton, J.A., Anderson, S.L., Van Voorhis, K., Maloy, C., and Siegel, E., 2002, Washington State marine water quality in 1998 through 2000: Washington State Department of Ecology, Environmental Assessment Program Publication 02-03-056.

State of Washington, variously dated, Marine water monitoring: Department of Ecology, State of Washington website, accessed March 1, 2008, at http://www.ecy.wa.gov/apps/eap/marinewq/mwdataset.asp.

Welschmeyer, N.A., 1994, Fluorometric analysis of chlorophyll a in the presence of chlorophyll *b* and phaeopigments: Limnology and Oceanography, v. 39, p. 1985-1992.

Wilde, F.D., ed., variously dated, National field manual for the collection of water-quality data—Field measurements: U.S. Geological Survey Techniques of Water-Resources Investigations, book 9, chap. A6, accessed July 21, 2008, at http://pubs.water.usgs.gov/twri9A6/.

World Ocean Circulation Experiment Hydrographic Program, 2008, Joint Global Ocean Flux Study (JGOFS) Protocols: Joint Global Ocean Flux Study website, accessed July 21, 2008, at http://usjgofs.whoi.edu/protocols_rpt_19.html.

Suggested citation

Lacy, J.R., and Dinicola, R.S., 2011, Aquatic environment—Circulation, water quality, and phytoplankton concentration, chap. 2 of Takesue, R.K., ed., Hydrography of and biogeochemical inputs to Liberty Bay, a small urban embayment in Puget Sound, Washington: U.S. Geological Survey Scientific Investigations Report 2011-5152, p. 9-40.

Chapter 2. Appendix A.   ADCP Data

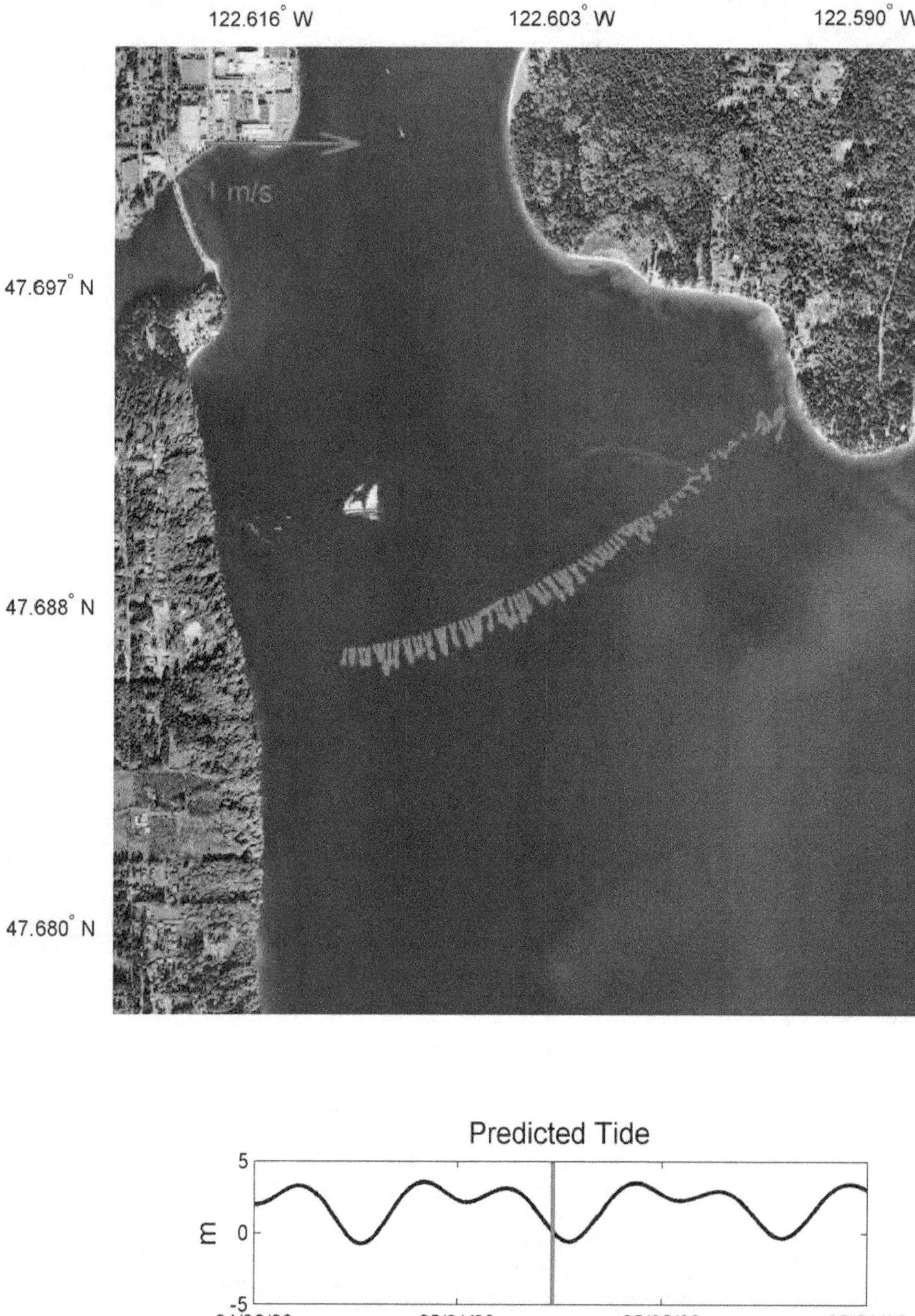

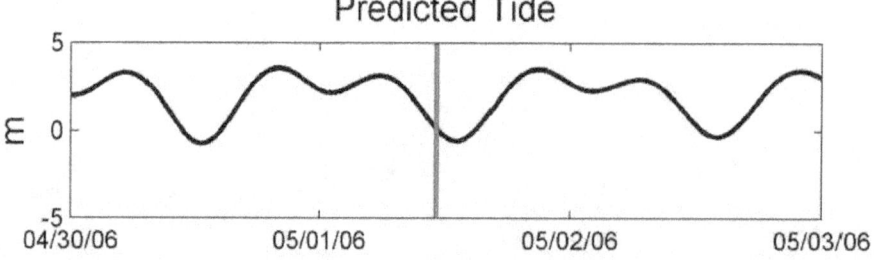

Figure A2-1.   Depth-averaged currents and tidal phase for ADCP transect 1.

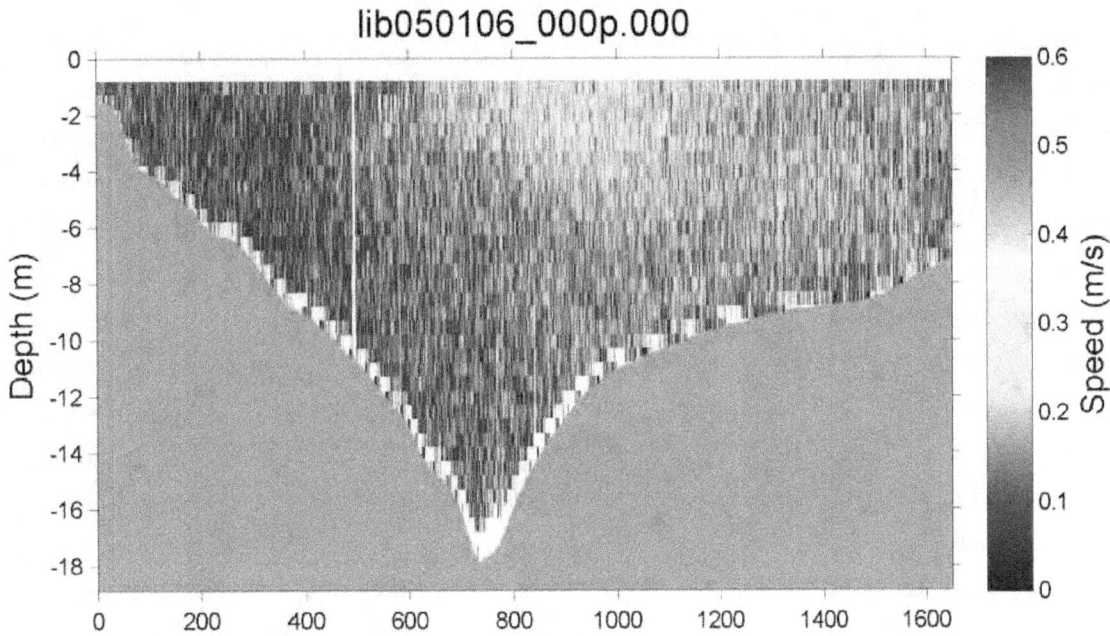

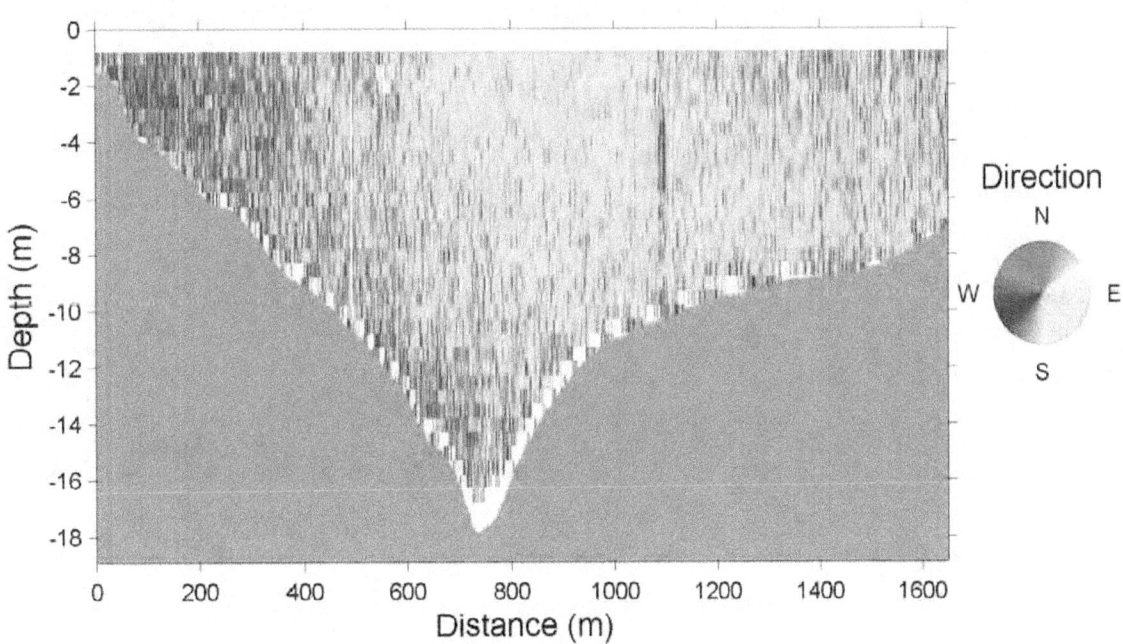

Figure A2-2.   Depth-dependent current speed (above) and direction (below) for ADCP transect 1.

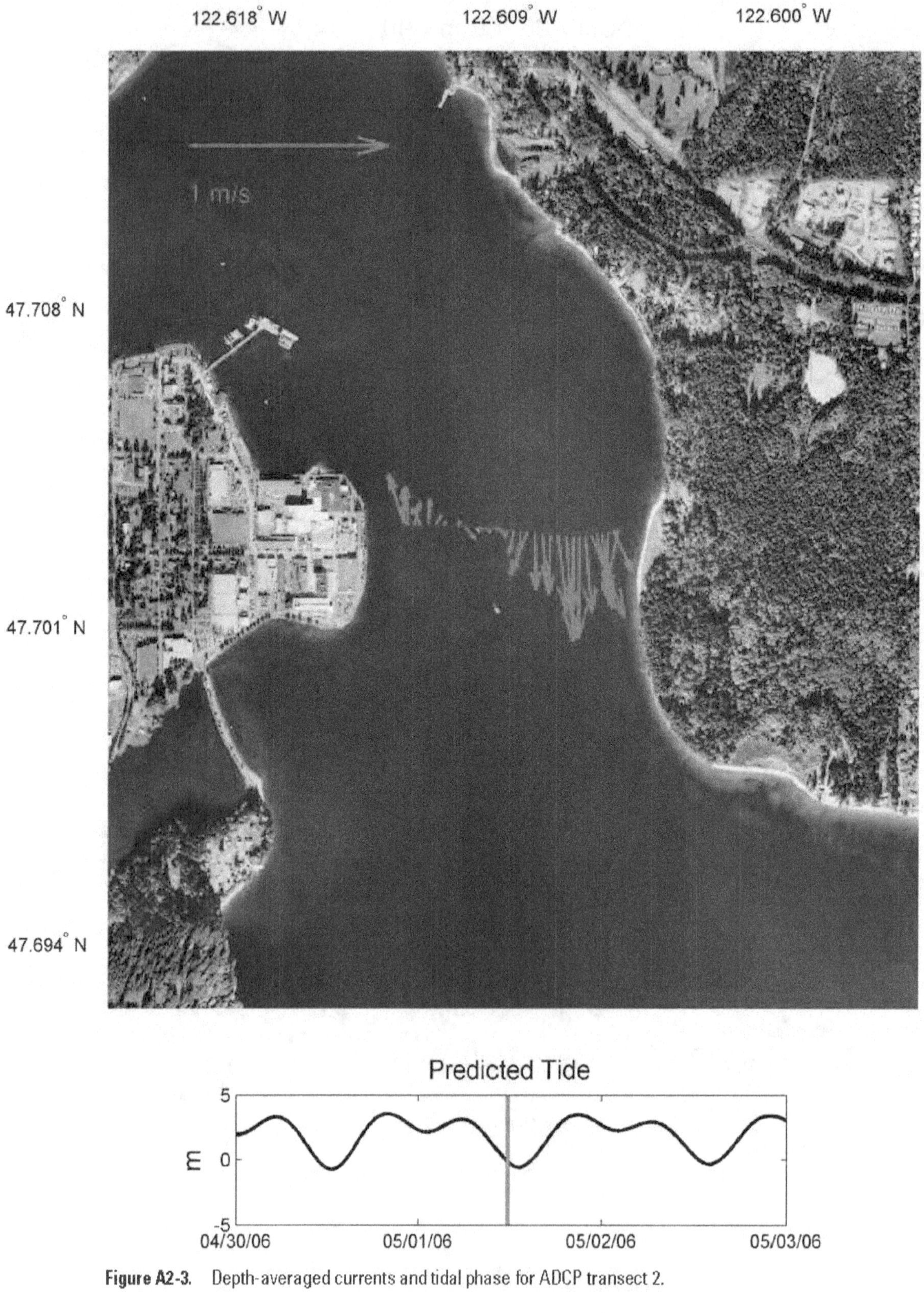

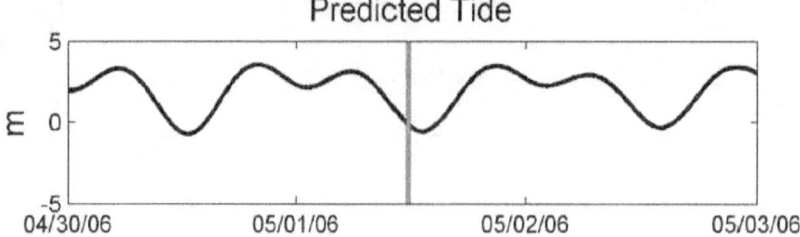

Figure A2-3. Depth-averaged currents and tidal phase for ADCP transect 2.

lib050106_001p.000

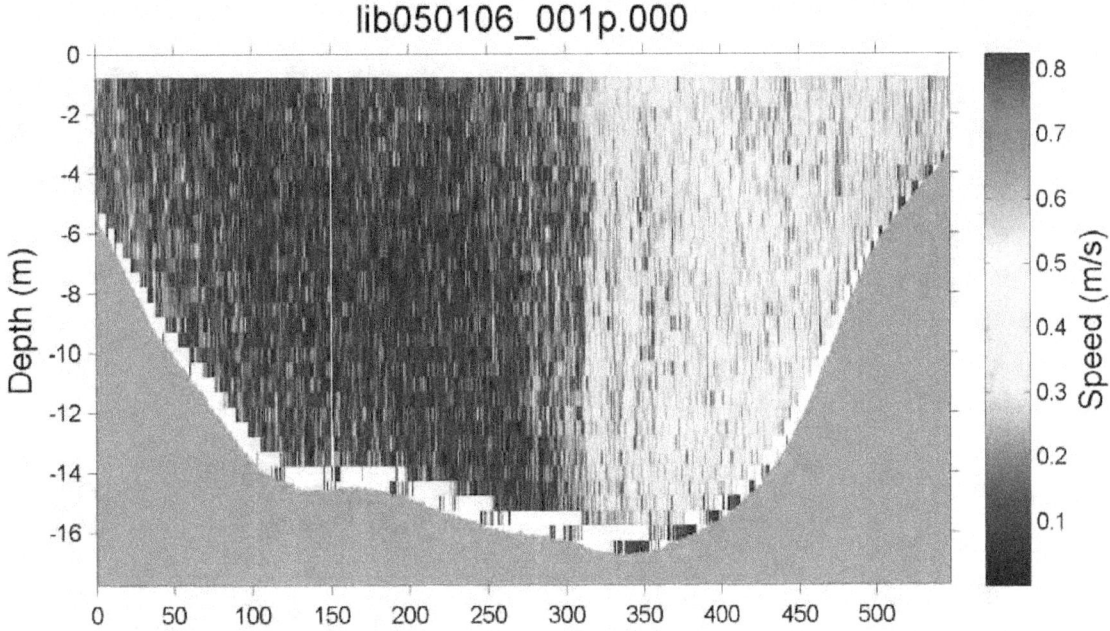

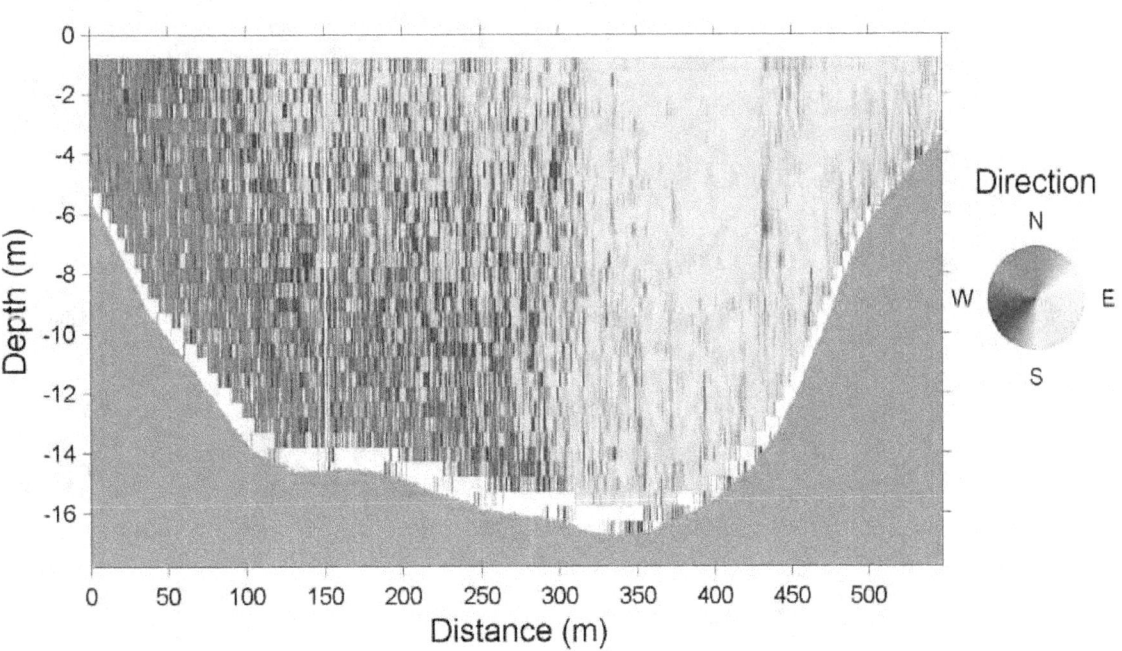

Figure A2-4. Depth-dependent current speed (above) and direction (below) for ADCP transect 2.

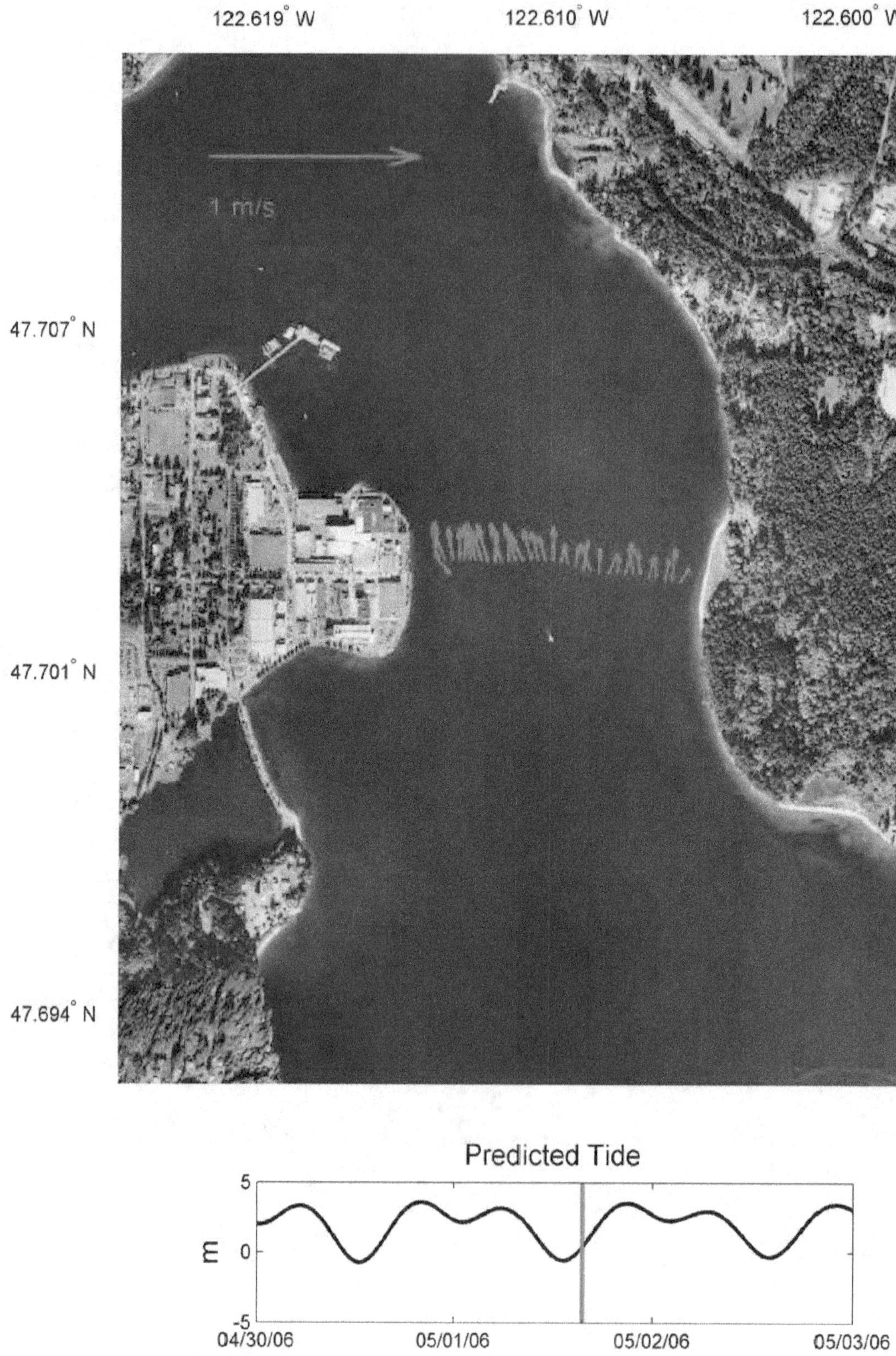

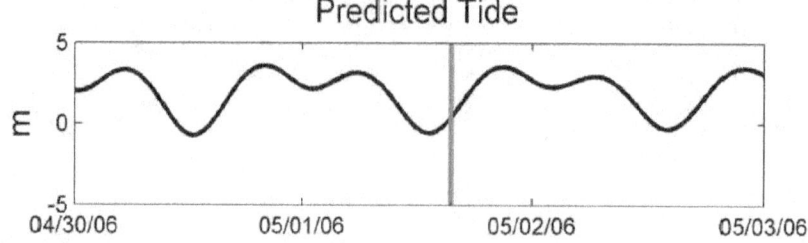

Figure A2-5.   Depth-averaged currents and tidal phase for ADCP transect 3.

lib050106_004p.000

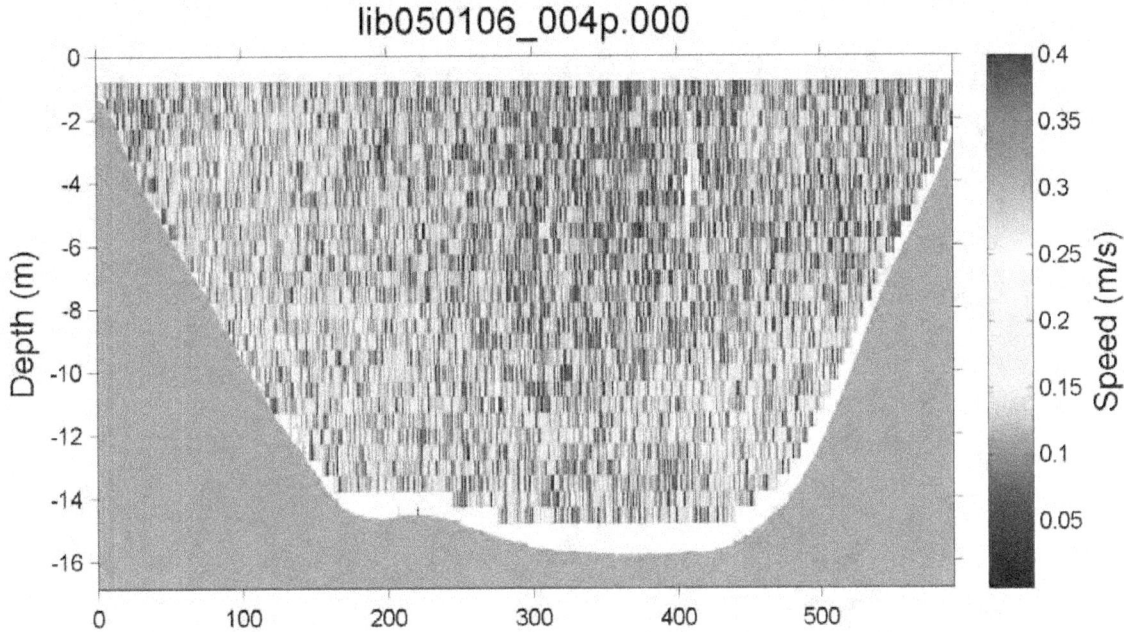

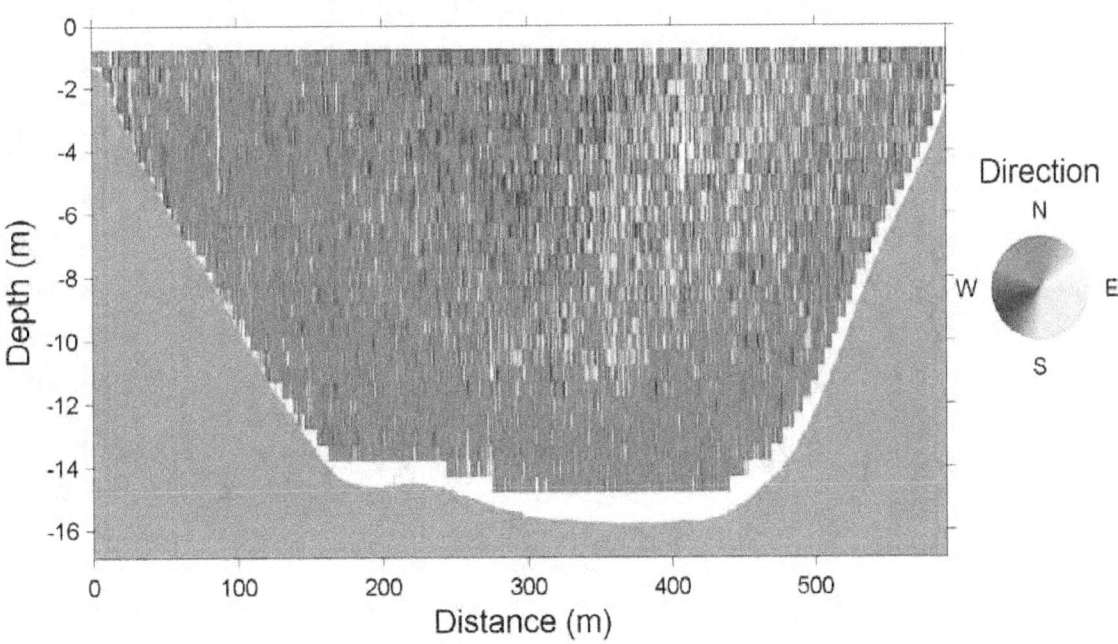

Figure A2-6.   Depth-dependent current speed (above) and direction (below) for ADCP transect 3.

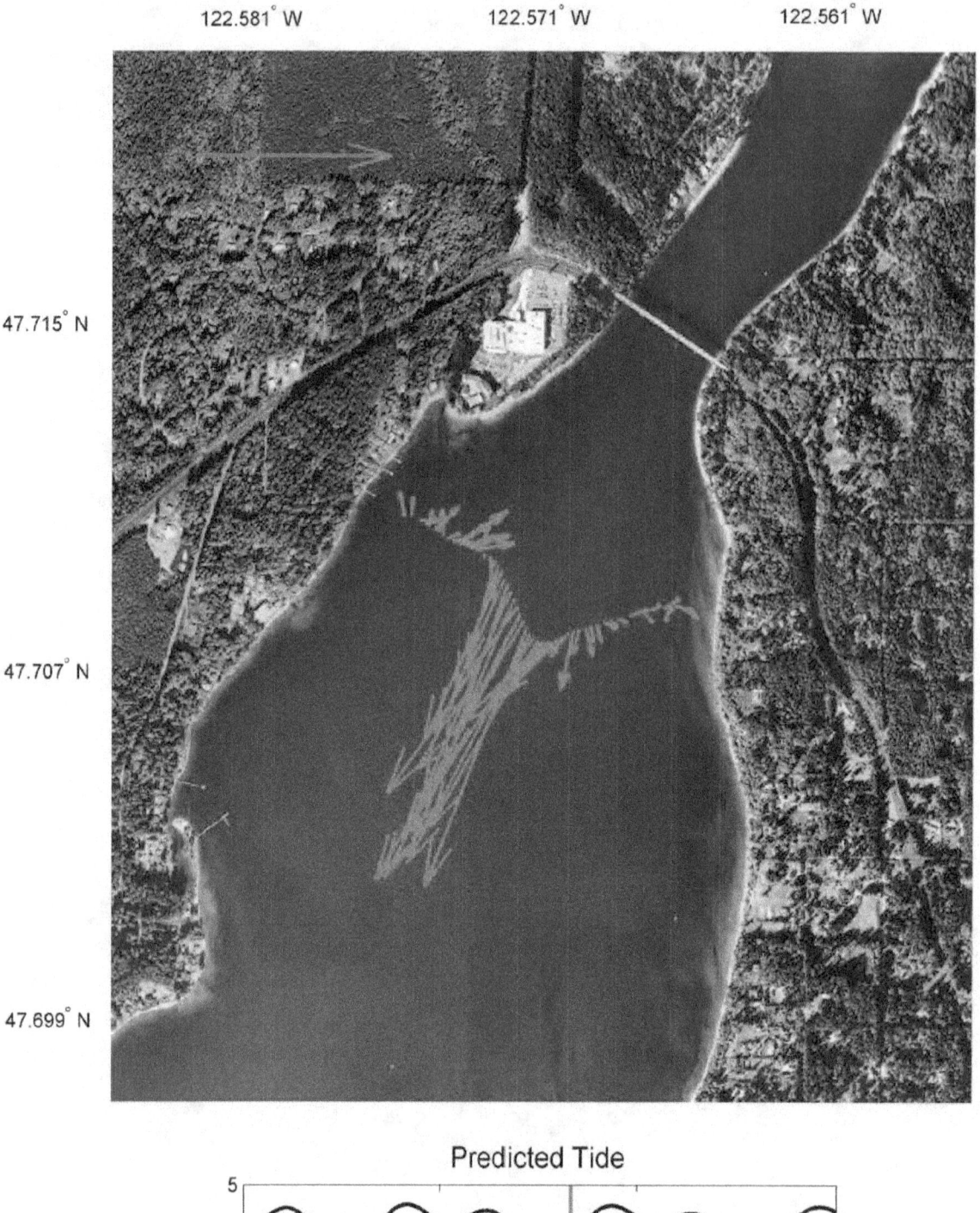

Figure A2-7.   Depth-averaged currents and tidal phase for ADCP transect 4.

lib050106_005p.000

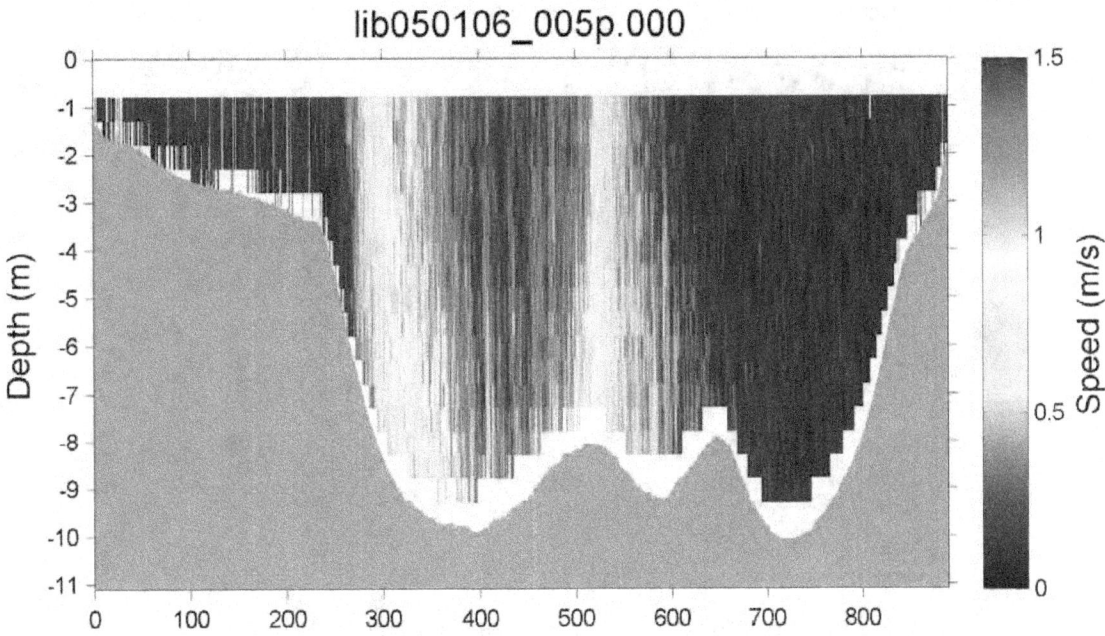

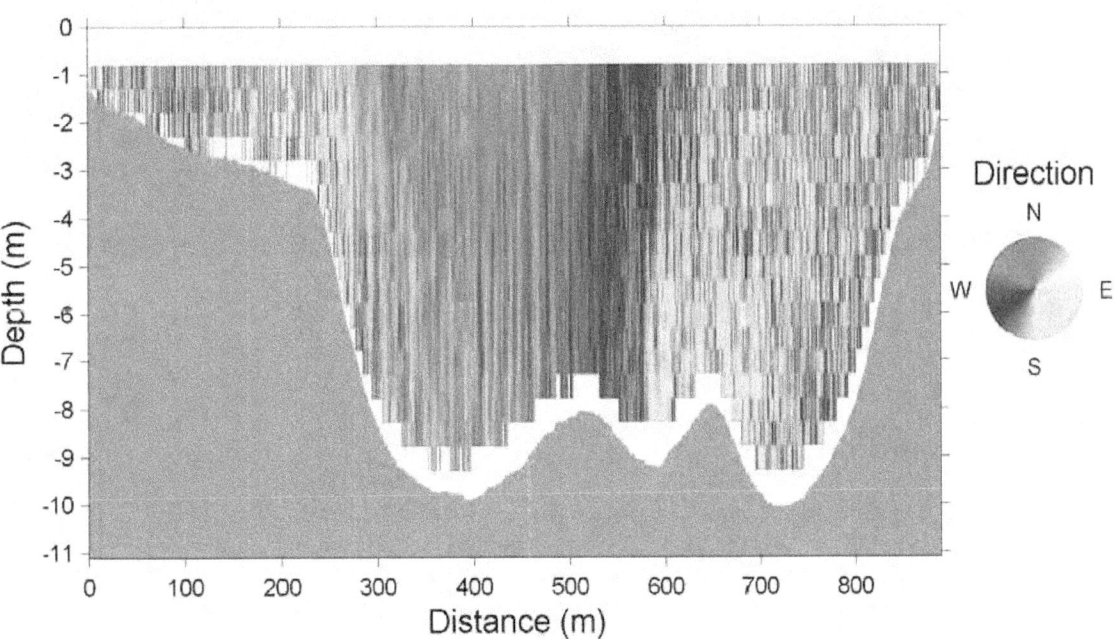

Figure A2-8.   Depth-dependent current speed (above) and direction (below) for ADCP transect 4.

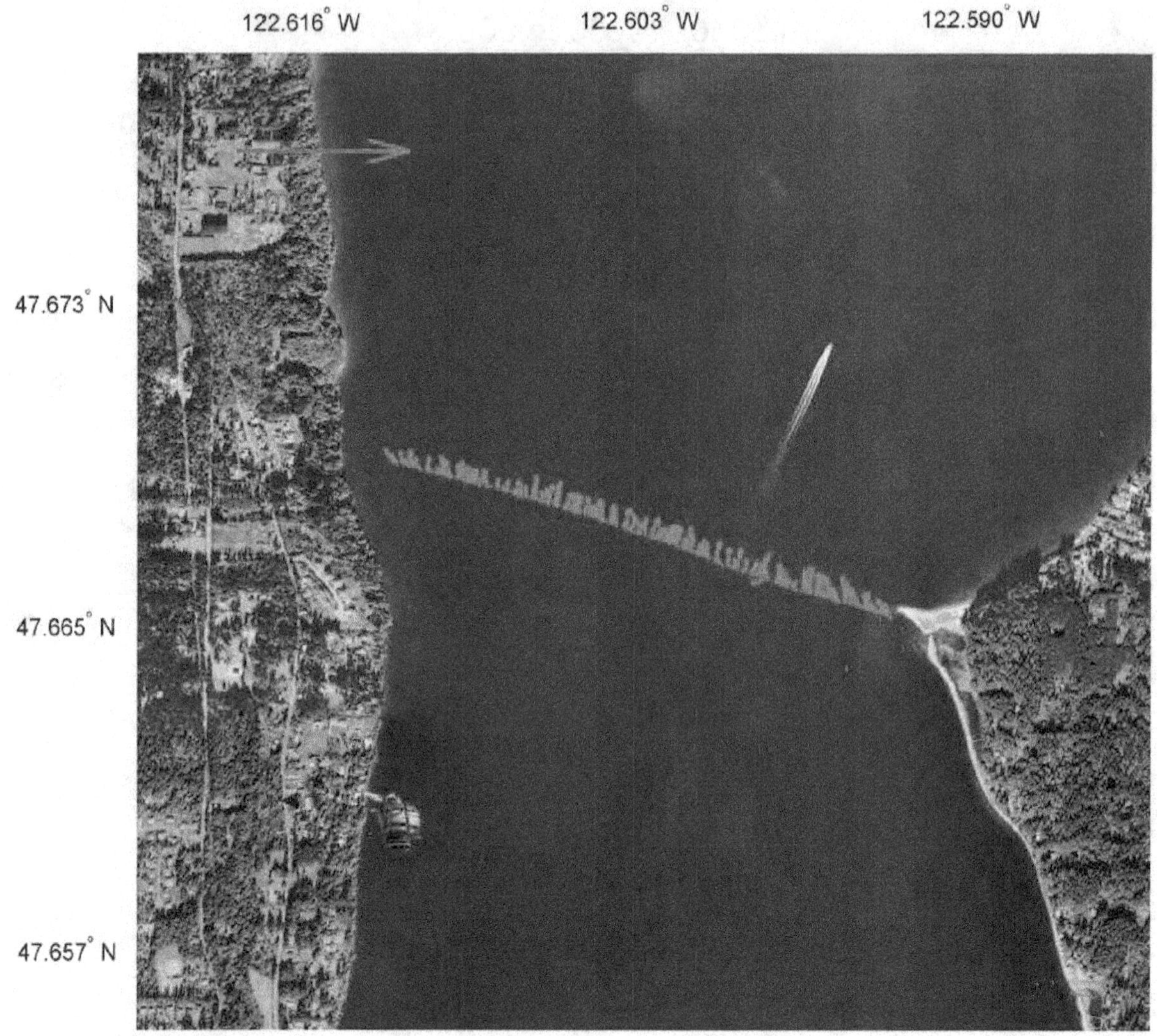

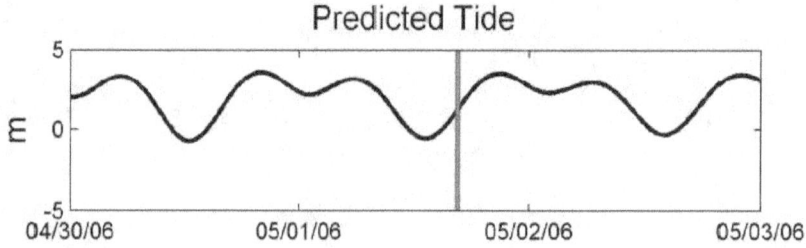

Figure A2-9.   Depth-averaged currents and tidal phase for ADCP transect 6.

lib050106_006p.000

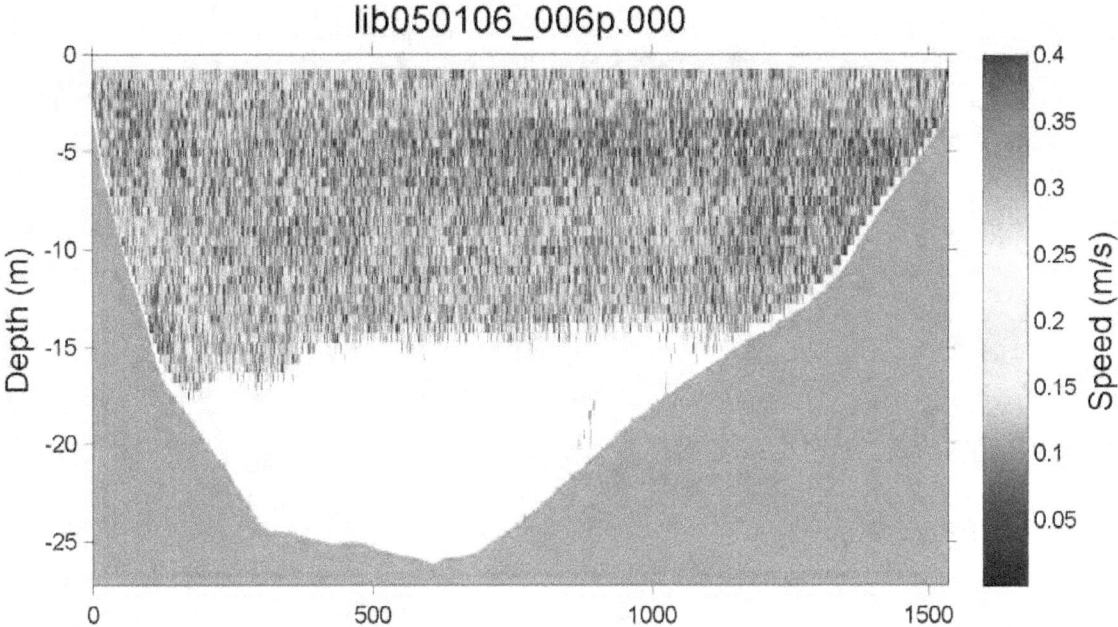

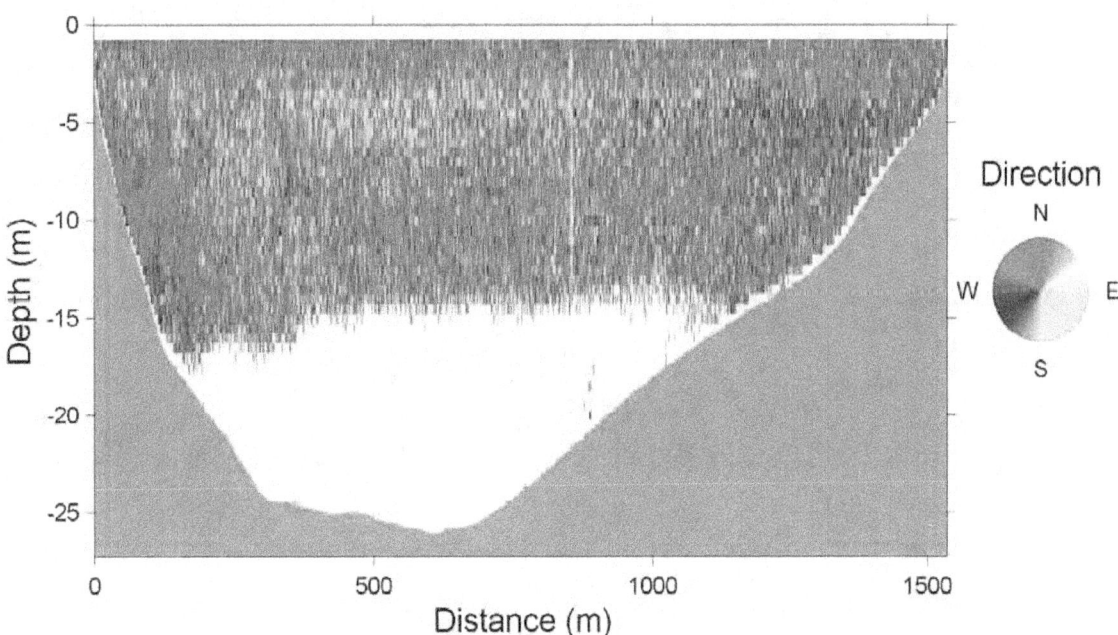

Figure A2-10. Depth-dependent current speed (above) and direction (below) for ADCP transect 5. Water depth was too great for the 1200 kHz ADCP to measure currents in the lower portion of the transect.

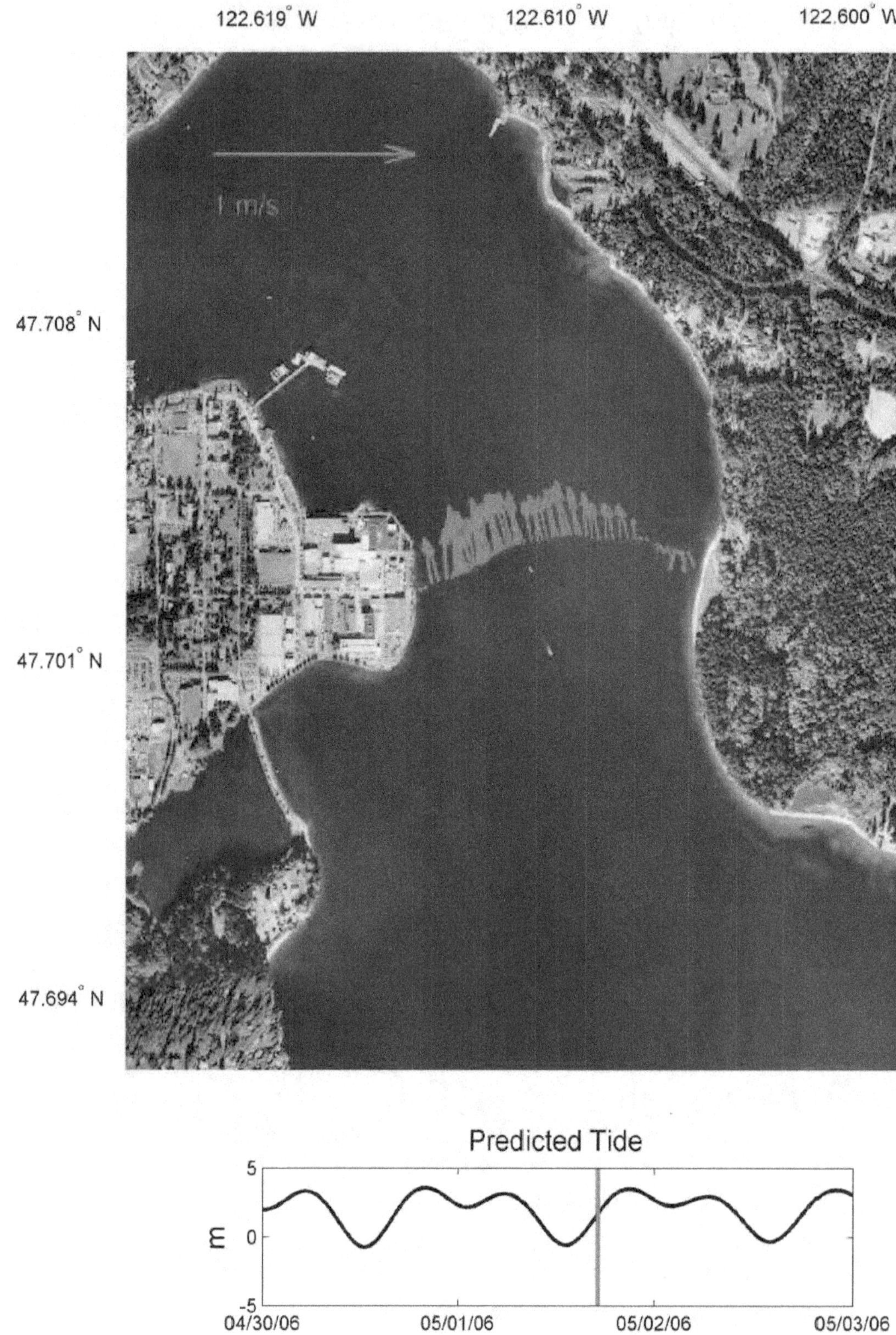

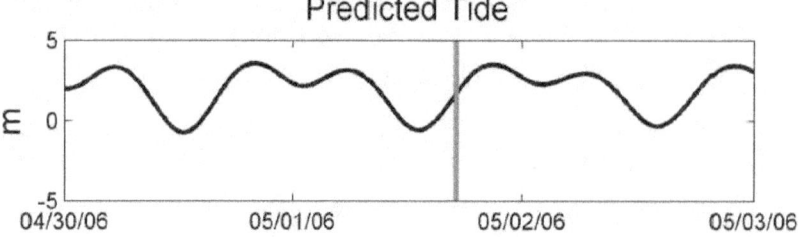

Figure A2-11.   Depth-averaged currents and tidal phase for ADCP transect 6.

lib050106_007p.000

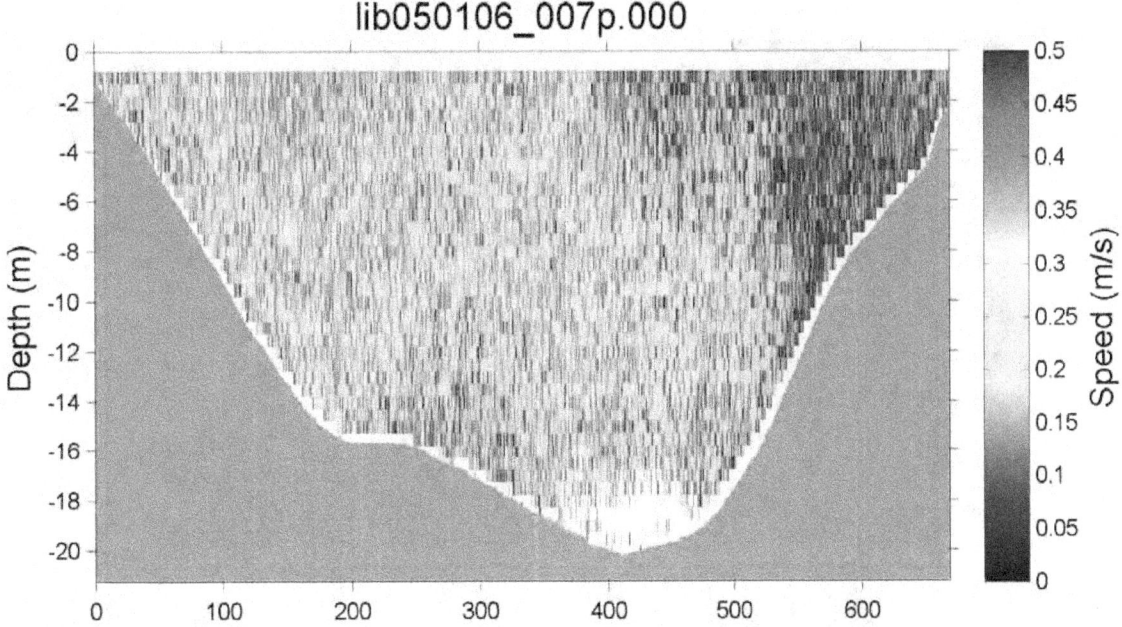

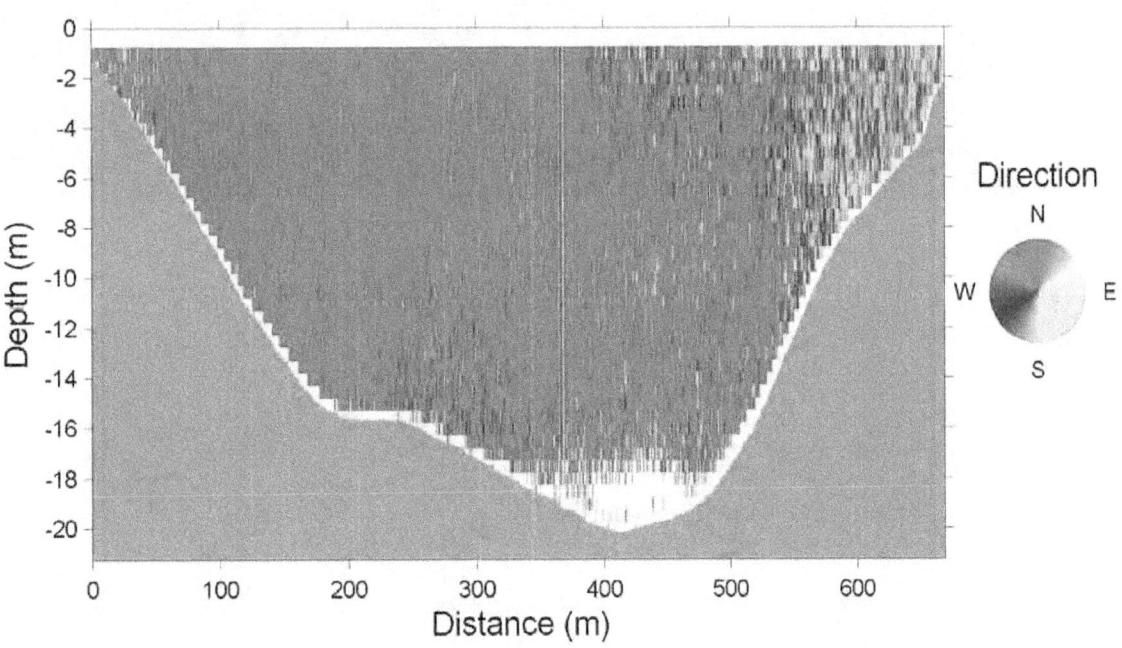

Figure A2-12. Depth-dependent current speed (above) and direction (below) for ADCP transect 6.

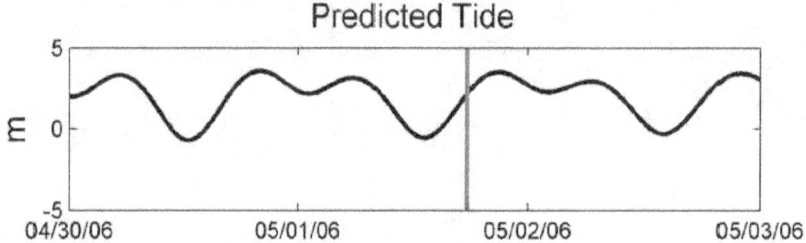

Figure A2-13.   Depth-averaged currents and tidal phase for ADCP transect 7.

lib050106_008p.000

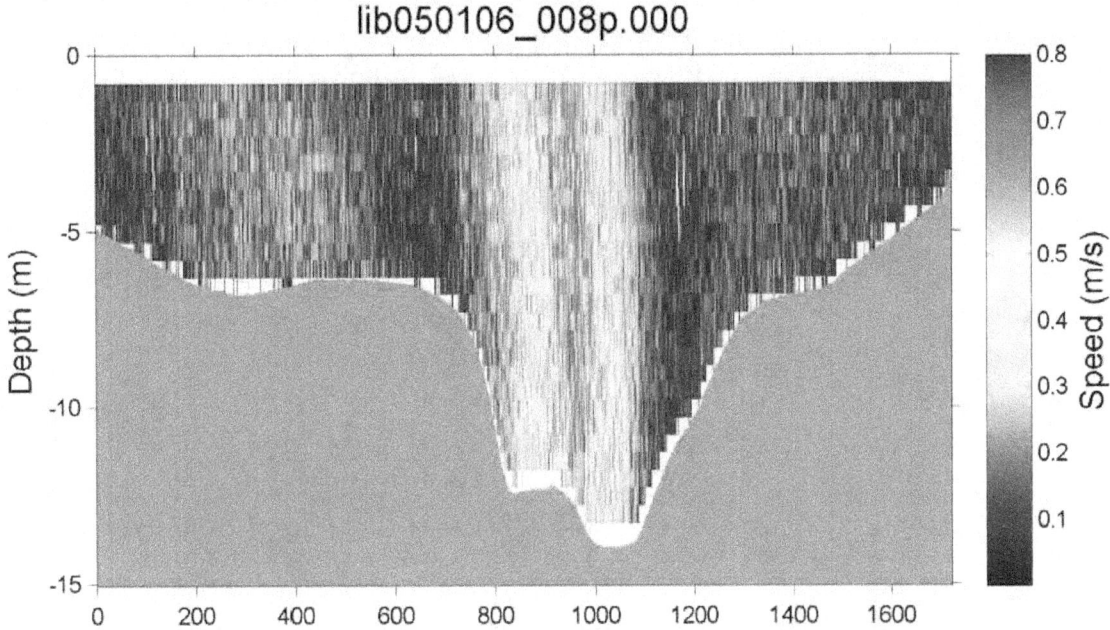

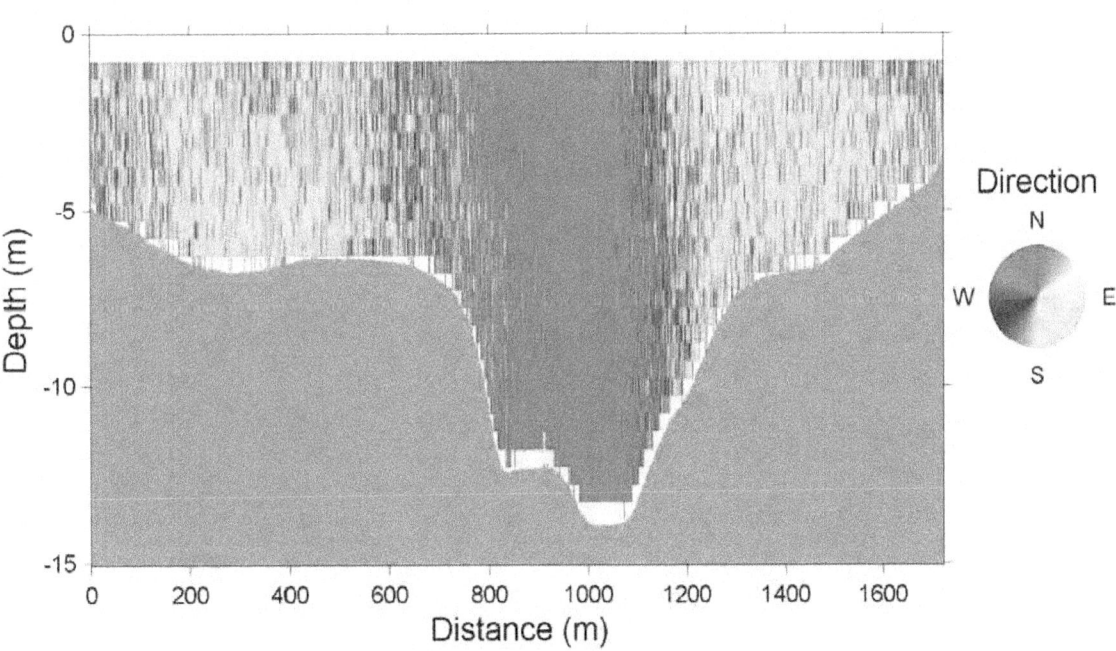

Figure A2-14. Depth-dependent current speed (above) and direction (below) for ADCP transect 7.

Chapter 3. Select Inorganic and Organic Loadings to Nearshore Liberty Bay, Puget Sound, Washington

By Richard S. Dinicola[1], Peter W. Swarzenski[2], and Jennifer Dougherty[3]

Introduction

Nearshore environments of Puget Sound increasingly are disturbed by human activities that can alter physical, chemical, and biological conditions and processes, and impair ecological functions. This chapter describes selected chemical inputs to nearshore waters that generally are associated with human disturbances in watersheds and that have not previously been examined in the study area. These inputs include nutrient loads in submarine groundwater discharge, and pharmaceutical and personal care product residues in coastal streams and groundwater.

Similar to other lines of investigation conducted during this study, a comparative approach between a semi-urbanized site (Liberty Bay) and a non-urbanized site (Point Bolin) was used to investigate potential differences in chemical inputs associated with a high degree of urbanization. The focus on a semi-urbanized site indicates the likelihood that population growth in the Puget Sound basin will lead to expanded residential development rather than to new large urban or industrial center developments.

Nutrient Loads in Submarine Groundwater Discharge

The highly productive, glacially derived coastal aquifers of the Puget Sound Regional Aquifer System are well known for their abundant water resources (Vaccaro and others, 1998) and their unique role in defining nearshore ecosystems (Swarzenski and others, 2007a). Groundwater often is the sole source of drinking water for coastal communities and is an important source of water for the aquatic ecosystems that define the landscape of Puget Sound. Previous investigations have shown the relation between increased urbanization and increased nutrient concentrations in groundwater in the Puget Sound region (Tesoriero and Voss, 1997). For this investigation, increased urbanization was investigated to determine if it also may lead to increased nutrient loads to the nearshore ecosystem through submarine groundwater discharge (SGD) processes. To provide a regional context, the estimated nutrient loads in SGD to Liberty Bay and vicinity were compared to nutrient loads similarly estimated from a site on Hood Canal with rural land use, and from sites on Skagit Bay that may be affected by nearby agricultural land-use.

Recent investigations have shown that the discharge of groundwater directly into coastal waters plays an important role in maintaining the biological diversity and productivity of many nearshore ecosystems, such as pocket-estuaries and wetlands. Investigations also have shown that nitrogen transported in coastal groundwater is an important component of the nutrient budget of New England and South Carolina salt marshes. In Great South Bay, New York, Bokuniewicz (1980) quantified SGD inputs, which were subsequently evaluated in terms of an important and substantial source of nitrate to the bay. From a similar study of SGD-derived nutrient fluxes into Florida Keys surface waters, Lapointe and others (1999) reported elevated nitrogen (N) and phosphorus (P) fluxes that may contribute to local phytoplankton blooms. In Tampa Bay, Swarzenski and others (2007b) quantified SGD rates using radium isotopes and then measured SGD-derived nutrient fluxes to the bay, which were at least on the same order of magnitude as riverine nutrient loading estimates. In the Loxahatchee River estuary of southeastern Florida, the role of SGD and SGD-derived nutrient fluxes was evaluated and compared to riverine estimates. The direct discharge of submarine spring water into ambient seawater caused a measurable dilution of salinity in Discovery Bay, Jamaica, and in the Atlantic Ocean, off northeastern peninsular Florida. Closer to Puget Sound, the estimate of nitrogen flux in SGD to the marine waters of Lynch Cove in Hood Canal, Washington, was one to two orders of magnitude larger than similar estimates derived from atmospheric deposition and surface water runoff, respectively (Swarzenski and others, 2007c).

---

[1] U.S. Geological Survey, Washington Water Science Center, 934 Broadway, Tacoma, WA 98402.

[2] U.S. Geological Survey, USGS Pacific Coastal and Marine Science Center, 400 Natural Bridges Drive, Santa Cruz, CA 95060.

[3] Department of Environmental Engineering, Stanford University, Stanford, CA 94305.

Submarine groundwater discharge is an almost ubiquitous coastal process that may affect nearshore material budgets (Caponne and Bautista, 1985). Although the contribution of SGD-derived nutrients, bacteria, carbon, and select trace elements such as barium or uranium (Charette and Sholkovitz, 2006; Swarzenski and Baskaran, 2007; Swarzenski, 2007) can vary widely depending on local hydrogeologic conditions and anthropogenic perturbations, accurate assessments of the spatial and temporal distribution of SGD along a particular coastline remain rare (Burnett and others, 2002; Dulaiova and others, 2006). There is little reliable data because SGD remains the 'hidden' vector in water and material transport from land to the sea and because the physical drivers of SGD are complex, often interrelated, and still poorly constrained (Taniguchi, 2002; Michael and others, 2005; Robinson and others, 2007b,c). Furthermore, the discharge of submarine groundwater usually is expressed not through well-defined marine springs (Swarzenski and others, 2001) but through diffuse discharge that can be ephemeral and spatially discontinuous (Burnett and others, 2002; Taniguchi and others, 2003).

Nutrient loads in submarine groundwater that originate from the Poulsbo area and from a less urbanized area northeast of Point Bolin were estimated in April 2007. The fluxes of silicate, nitrate and nitrite, ammonia, and total organic nitrogen were estimated at relatively urbanized Oyster Plant Park in Poulsbo and at less urbanized Sandy Hook northeast of Point Bolin (fig. 3-1). Associated submarine groundwater discharge estimates were derived using seepage meters, geochemical tracers, and electrical resistivity methods.

Theory and Methods for Measuring Submarine Groundwater Discharge

One of the simpler and widely used devices to measure direct fluid exchange rates across the sediment/water interface is the manual seepage meter. Second generation electromagnetic (EM) seepage meters function autonomously and measure bidirectional fluid exchange rates, and thus provide much more subtle information on the response of fluid exchange rates to external forcing, such as tides and waves and terrestrial precipitation/recharge. A shortcoming of any seepage meter is the small footprint of the instrument that at best can provide only site-specific information for fluid exchange. Regardless, one EM seepage meter was used at each SGD study site for this investigation (fig. 3-1).

In contrast to measurements by seepage meters, local to regional scale SGD information can be obtained from certain naturally occurring isotopes of the uranium/thorium (U/Th) decay series. The application of selected U/Th series radionuclides as unique tracers of SGD has developed along two contrasting themes: (1) the excess activity of a radionuclide in a coastal water body may be geochemically linked to groundwater discharge and (2) vertical pore water and solid phase activities are assessed within the constraints of an advection/diffusion model. To develop a mass balance model for submarine groundwater discharge, time series of excess radon isotope ^{222}Rn in nearshore surface water was utilized (Burnett and others, 2001, 2003, 2006; Burnett and Dulaiova, 2003; Dulaiova and others, 2005, 2006, Swarzenski and others, 2006; Weinstein and others, 2007). The premise of this technique relies on converting near-continuous excess ^{222}Rn (accounting for a mean parent radium isotope ^{226}Ra water column activity) measurements (disintegrations per minute per cubic meter, dpm m^{-3}) into inventories (disintegrations per minute per square meter, dpm m^{-2}) using real-time water-level data. Radon inventories are subsequently converted into hourly fluxes (dpm m^{-2} hr^{-1}) and then corrected for both tidal radon exchanges (positive as offshore ^{222}Rn brought in by an inflowing tide and negative as ^{222}Rn lost during an outflowing tide) and atmospheric ^{222}Rn losses. In order to convert nearshore total ^{222}Rn fluxes (dpm/m^2/hr) into advective exchange rates (cubic centimeters per square centimeter per day, cm^3 cm^{-2} d^{-1}), a representative groundwater end-member activity must be quantified. Groundwater ^{222}Rn time-series was acquired at both SGD study sites from shallow wells installed close to the high-tide line and instrumented for continuous water levels and specific conductivities (fig. 3-1).

The use of electrical resistivity to examine the dynamics of the fresh water/salt water interface in coastal aquifers is well established (Manheim and others, 2004) and has been recently enhanced by improvements in streamer configuration, data acquisition, and processing firmware and software (Swarzenski and others, 2006, 2007a). Electrical resistivity measurements map the fresh water/salt water interface because fresh water is generally more resistive (resistivity = specific conductivity^{-1}) as a result of relatively low concentrations of dissolved salts, whereas salt water has substantially lower resistivity as a result of relatively high concentrations of dissolved salts. For this study, stationary resistivity surveys were conducted using a cable with 56 electrodes spaced 2 m apart at each SGD study site to evaluate the width of the SGD zone. In stationary mode, an Advanced Geosciences, Inc. (AGI) external switching box connected to an R8 SuperSting multi-channel receiver controlled the current flow along the 56-electrode cable. The 112-m stationary cable was oriented perpendicular to the shore, and each electrode was pinned to the underlying sediment by a stainless steel 35-cm spike. The relative elevation of each electrode was measured using a laser level and the topography/bathymetry data were incorporated into inverse modeling routines (AGI EarthImager).

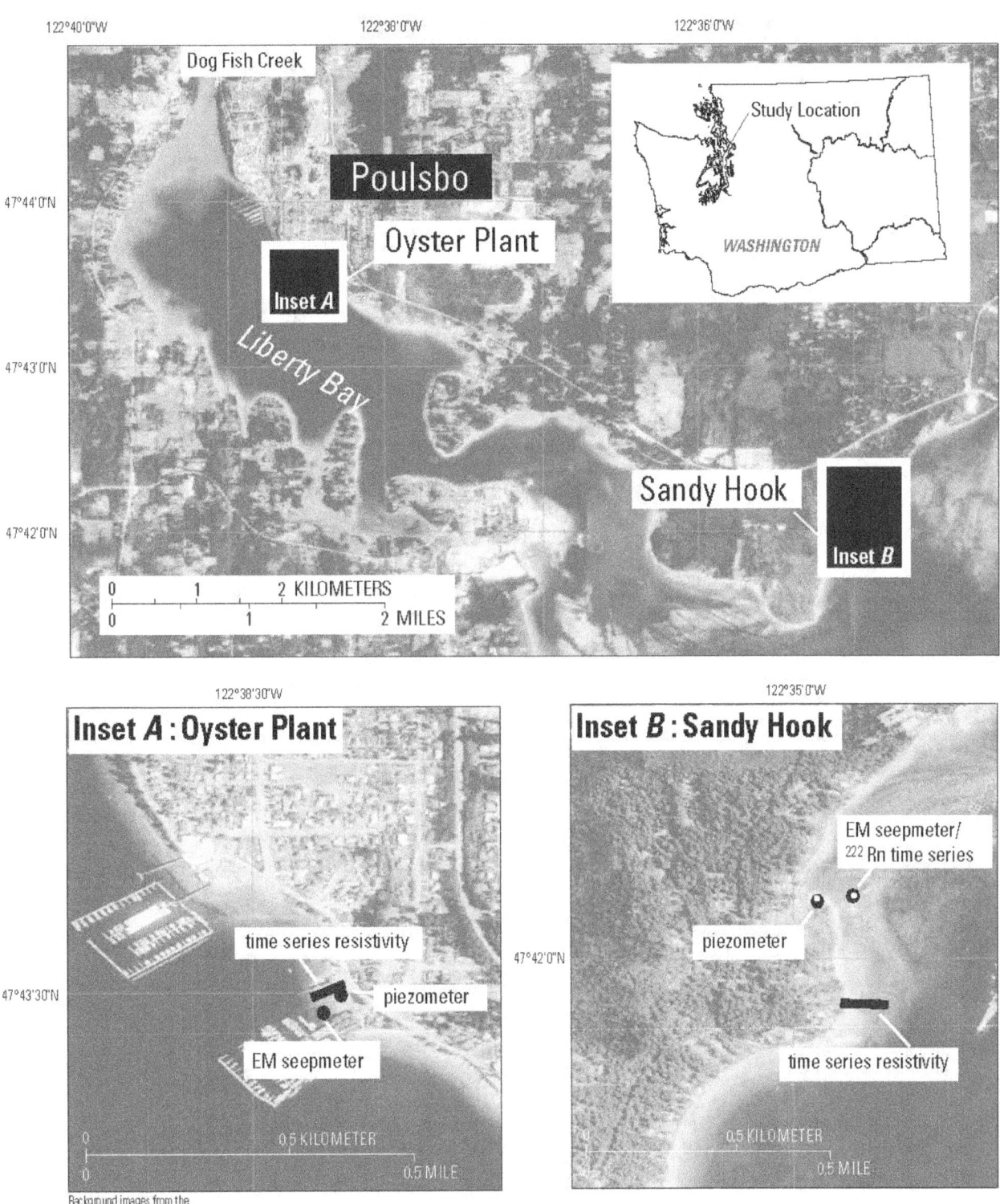

Figure 3-1.   Location of the submarine groundwater discharge study sites, Kitsap County, Washington.

Submarine Groundwater Discharge and Associated Nutrient Loads

Field observations of time-series ^{222}Rn and EM seepage meter measurements at Oyster Plant Park and Sandy Hook (table 3-1) reveal the average SGD rates for these two sites were similar in magnitude, but were about four times less than rates measured at Lynch Cove in Hood Canal. The stationary resistivity lines (fig. 3-2) suggest that the width of the zone under the effect of active SGD was similar for the Lynch Cove and Liberty Bay sites; ranging 25–31 m.

To estimate nutrient loads in SGD, nutrient concentrations were determined for groundwater samples collected from shallow piezometers (approximately 2 m deep) at Oyster Plant Park and Sandy Hook (fig. 3-1). Approximately 100 mL of groundwater was collected in opaque nalgene bottles from the piezometers using a peristaltic pump and polyethylene tubing. The samples were stored on ice and sent to Woods Hole Oceanographic Institute for analysis of a suite of nutrients (table 3-2) using standard colorimetric techniques (http://www.whoi.edu/sbl/liteSite.do?litesiteid=1671&articleId=2922, last accessed April 2, 2011).

Consistent with previously described relations between urbanization and nutrients in groundwater, the DIN concentration (sum of the concentrations of nitrate, nitrite, and ammonia) measured in shallow groundwater (about 4 m below ground surface) at relatively urban Oyster Plant Park was substantially greater than the DIN concentration measured in groundwater at relatively rural Sandy Hook (table 3-2). Similarly, the DIN concentration measured in nearshore seawater at Oyster Plant Park was greater than the DIN concentration measured at Sandy Hook. Because the DIN concentrations in groundwater at Oyster Plant Park and Sandy Hook were four and two times greater, respectively, than the DIN concentrations measured in nearshore seawater, groundwater discharge to seawater potentially may increase nearby marine nitrogen concentrations. However, nutrient concentrations in marine waters in of the study area are highly variable. The total dissolved nitrogen (TDN) concentration measured in nearshore marine water at Oyster Plant Park in April 2007 (1,430 µg L^{-1}) was much greater than the highest TDN concentration (240.8 µg L^{-1}) measured in Liberty Bay during April–May 2006 (table 2-1). In contrast, the TDN concentration measured in nearshore marine water at the Sandy Hook site in April 2007 (299 µg L^{-1}), was similar to the highest TDN concentration (220.2 µg L^{-1}) measured at Point Bolin in April–May 2006. Regardless of differences in TDN concentrations, these data do indicate that groundwater is a source of nitrogen for the marine waters of the study area.

The calculated SGD rates (table 3-1) were multiplied by the measured concentrations of nitrogen compounds in groundwater (table 3-2) to estimate nutrient flux rates for the Liberty Bay sites (table 3-3). The measured DIN concentrations resulted in estimated fluxes of DIN in SGD from Oyster Plant Park of 0.3 moles per day per meter of shoreline (mol d^{-1} m^{-1}). The DIN fluxes from Oyster Plant Park are more than three times greater than the estimated DIN flux from less urbanized Sandy Hook (0.1 mol d^{-1} m^{-1}) despite a larger freshwater SGD rate for the less urbanized site. The estimated total dissolved nitrogen (TDN) flux rates for

Table 3-1. Summary of mean submarine groundwater discharge rates for Liberty Bay, Lynch Cove, Hood Canal, as calculated from stationary ^{222}Rn time series deployments and electromagnetic seepage meters.

[Width of the discharge zone was delineated by land-based, multi-channel resistivity surveys. Average discharge rates depicts upward groundwater flow, per meter of shoreline. Lynch Cove data as reported in Swarzenski and others, 2007. **Abbreviations:** m, meter; cm d^{-1}, centimeter per day; m^3 d^{-1} m^{-1}; cubic meter per day per meter; –, not analyzed; ≤, less than or equal to]

Study area	Site name	Submarine groundwater discharge (SGD) rate		Width of SGD zone (m)	Average SGD rate per meter of shoreline (m^3 d^{-1} m^{-1})
		^{222}Rn time series	Electromagnetic seepage meter		
		(cm d^{-1})			
Liberty Bay	Oyster Plant	–	19.3±15.2	25	4.8
	Sandy Hook	27.8±11.5	11.6±5.5	31	6.1
Lynch Cove	Merrimont Dock	85±84	≤81	26	22

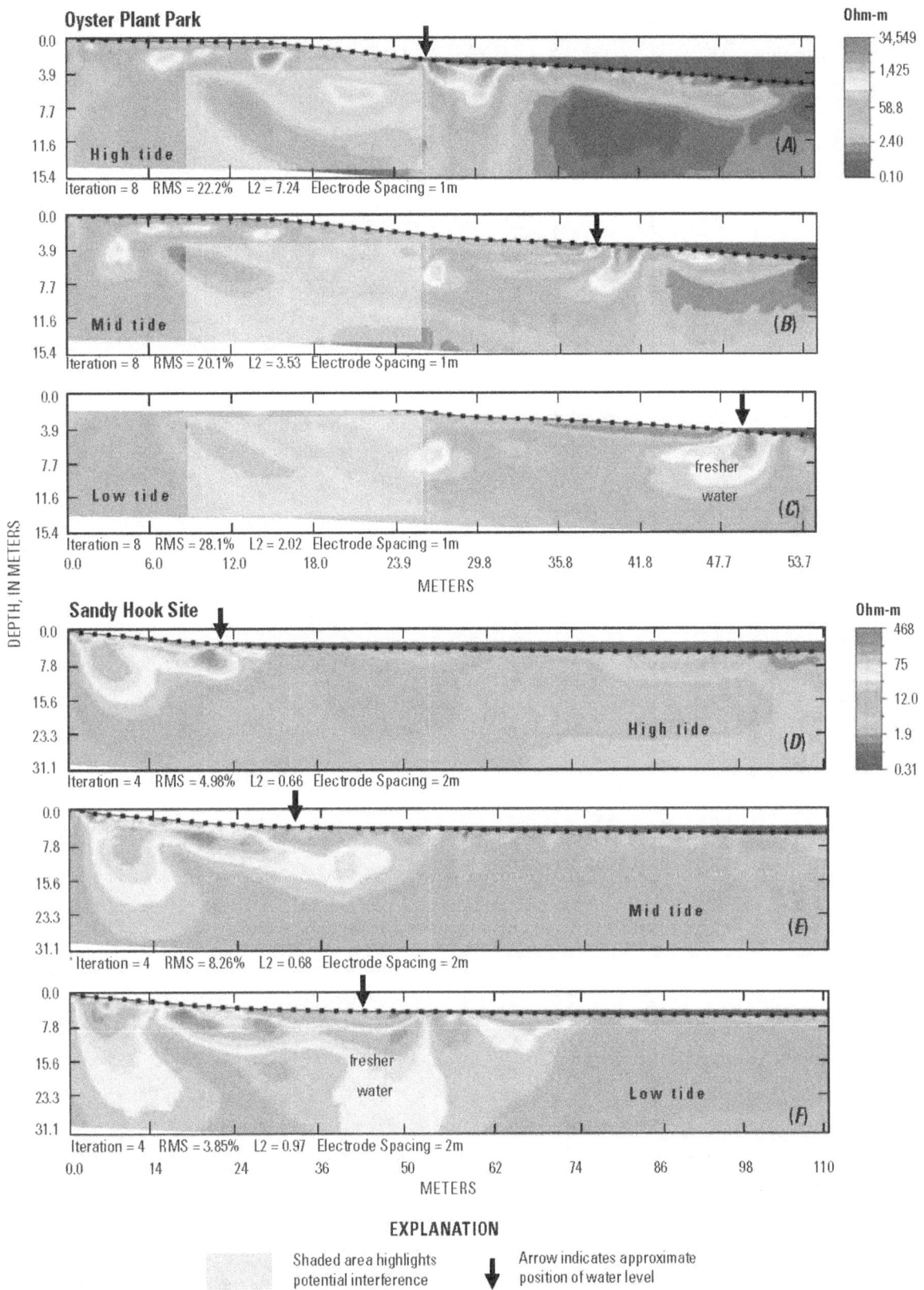

Figure 3-2. Electrical-resistivity profiles perpendicular to the shoreline at the Oyster Plant Park and Sandy Hook sites, Liberty Bay and vicinity, Puget Sound, Washington. Each profile represents a snapshot of electrical resistivity along a cross-section of the intertidal area, as indicated in figure 3-1. Colors represent resistance in ohm-meters. Red colors indicate more electrically resistant fresh water, and blue colors indicate more electrically conductive saline water. Approximate position of the waterline on the beach is indicated by the arrow above the profile.

Table 3-2. Average nutrient concentrations in groundwater and adjacent nearshore seawater at Oyster Plant Park and Sandy Hook, Kitsap County, Washington.

[Abbreviations: n, number of samples analyzed; µg/L, microgram per liter; psu, practical salinity units]

Sample type	Date	Dissolved phosphate (µg/L as P)	Dissolved silica (µg/L as Si)	Dissolved nitrite+nitrate (µg/L as N)	Dissolved ammonia (µg/L as N)	Dissolved inorganic nitrogen (µg/L as N)	Dissolved organic nitrogen (µg/L as N)	Total dissolved nitrogen (µg/L as N)	Salinity (psu)
				Oyster Plant Park					
Groundwater (n=1)	04-24-07	41	8,360	914	0.98	915	516	1,430	<0.2
Nearshore seawater (n=1)	04-24-07	2.2	963	74.3	96.3	171	1,430	1,600	2.7
				Sandy Hook					
Groundwater (n=3)	04-24-26 2007	105	1,007	1.12	165	165	195	360	0.3
Nearshore seawater (n=3)	04-24-26 2007	38.5	1,376	54.4	26.9	81.3	218	299	19.6

Table 3-3. Combined submarine groundwater discharge-derived nutrient loading estimates (mol d⁻¹ per m of shore line) into the nearshore coastal waters at Liberty Bay and Hood Canal, Washington.

[Estimates were made using the average Submarine Groundwater Discharge (SGD) rates shown on table 3-1. **Abbreviations**: mol d^{-1} m^{-1}, moles per day per meter of shoreline]

Study area	Site name	Dissolved ammonia	Dissolved silica	Dissolved nitrite + nitrate	Dissolved inorganic nitrogen (DIN)	Dissolved organic nitrogen (DON)	Total dissolved nitrogen (TDN)
					(mol d^{-1} m^{-1})		
Liberty Bay	Oyster Plant	0	1.4	0.3	0.3	0.2	0.5
	Sandy Hook	0.1	0.2	0	0.1	0.1	0.2
Lynch Cove	Merrimont Dock	0	3.2	0.5	0.5	0.1	0.6

Oyster Plant Park and Sandy Hook were 0.5 and 0.2 mol d^{-1} m^{-1}, respectively (TDN includes dissolved organic nitrogen-bearing compounds in addition to DIN). The DIN and TDN flux estimates for Oyster Plant Park and Sandy Hook were slightly less than similarly derived DIN and TDN fluxes (0.5 and 0.6 mol d^{-1} m^{-1}, respectively) estimated at the Merrimont Dock in Lynch Cove, Hood Canal (Swarzenski and others, 2007c) where nearby land use is similar to that at Sandy Hook. However, the DIN and TDN flux rates at all three sites were substantially less than the DIN flux rates of 1.8 to 3.2 mol d^{-1} m^{-1}, and the TDN flux rates of 2.0 to 6.5 mol d^{-1} m^{-1} estimated at three sites in Skagit Bay that may have been affected by nearby agricultural land uses (P.W. Swarzenski, written commun., February 8, 2009). Overall, these data indicate that groundwater is a source of nitrogen for the marine waters of the study area.

Pharmaceutical and Personal Care Product Residues in Surface Water and Groundwater

In recent years, the presence of so-called emerging contaminants such as pharmaceuticals and personal care products (PPCPs) in our surface and ground waters has received greater attention (Kolpin and others, 2002; Glassmeyer and others, 2005). The sources of this contamination are varied, and include wastewater treatment plants (WWTPs) and decentralized septic systems (Geary, 2005; Hinkle and others, 2005; Conn and others, 2006; Swartz and others, 2006; Godfrey and others, 2007). The question whether treatment in septic systems is adequate to protect groundwater and surface water from PPCP contamination has not been adequately addressed. This question is important because Moore (1996) has shown that the contribution of groundwater into coastal water can be large.

The following is a summary of methods and results from a companion study (Dougherty and others, 2010) that examined the occurrence of selected emerging contaminants (herbicides, PCPPs, and a flame retardant) in groundwater and surface water for the areas near Poulsbo that primarily are served by on-site waste systems. Additional details and discussion of the methods and results are available in Dougherty and others (2010). The contaminants analyzed included a broad spectrum of 25 commonly used compounds that were selected in part based on the analytical method adapted from Vanderford and others (2003) that allowed for analysis of multiple compounds with one method. The compounds were analyzed in surface-water samples from eight creeks, shallow groundwater samples collected from three sites around Liberty Bay and Sandy Hook, and passive integrative samplers deployed from January through March 2007 and again in July through September 2007 at two sites in lower Dogfish Creek, the main creek flowing into Liberty Bay. The PPCPs in particular are considered indicators of domestic wastewater and in our study area most likely originate from local septic systems.

Sampling and Analysis Methods

Discrete surface-water grab samples were collected from eight creeks flowing into Liberty Bay in spring 2007. Two liters of water were collected in two 1-L amber glass bottles (I-CHEM; Rockwood, TN, USA). Water samples were stored at 4 °C and processed within 48 hours of arrival at the Stanford University laboratory for analysis of the contaminants (Dougherty and others, 2010). Contaminant levels in discrete surface-water samples were expected to be very low based on the community size and non-point source nature of contaminant discharge.

A passive Polar Organic Chemical Integrative Sampler (POCIS; Environmental Sampling Technologies) also was used to concentrate target compounds over a long period, making it possible to detect target compounds that may be missed with traditional surface-water grab samples because of concentrations less than the method detection levels. Two types of POCIS were deployed in 2007 that separately targeted pharmaceutical and pesticide compounds (Alvarez and others, 2004) at the mouth of Dogfish Creek and 1 mi upstream during the wet season (January–March) for 62 days and during the dry season (July through September) for 61 days. Samplers were deployed and retrieved at low tide. The samplers were stored in methanol-rinsed cans at –20 °C and processed within 14 days of recovery (Dougherty and others, 2010).

Groundwater samples were collected from shallow piezometers (approximately 4 m deep) at Oyster Plant Park and Sandy Hook (fig. 3-1). One liter of groundwater was collected in a 1-L amber glass bottle (I-CHEM; Rockwood, TN, USA) from each piezometer using a peristaltic pump and new polyethylene tubing. The samples were stored on ice or at 4 °C and processed within 7 days of arrival at the Stanford University laboratory for analysis of contaminants (Dougherty and others, 2010).

Results

Selected emerging contaminants were detected in discrete and integrated surface-water samples in creeks flowing into Liberty Bay, as well as in groundwater (table 3-4). At least one contaminant was positively detected in samples from 8 of the 11 sites sampled. Of the 25 contaminants analyzed, 12 were detected at concentrations greater than the various levels of quantification that ranged between 0.5 and 10 ng/L or nanograms per POCIS filter (ng/filter). Concentrations of the 12 detected compounds ranged from 0.7 to 19 ng/L in discrete groundwater and surface water samples and 0.5 to 57 ng/filter in integrated samples. Average contaminant concentrations were 6 ng/L for surface-water grab samples, 8 ng/filter for surface-water passive samples, and 8 ng/L for groundwater samples. Contaminant concentrations in surface and groundwater samples cannot be quantitatively compared to concentrations in POCIS because they represent different sampling periods.

Table 3-4. Summary of emerging contaminant compounds that were positively detected in the vicinity of Liberty Bay, Washington during 2007.

[Compounds that were positively detected in one or more samples are in **bold**. The maximum possible number of detections was 8 for POCIS, 3 for GW, and 8 for SW. **Abbreviations:** DEET, N,N-diethyl-meta-toluamide; TCEP, tri(2-chloroethyl) phosphate; POCIS; polar organic chemical integrative sampler; GW, groundwater; SW, surface water]

Compound	Class	Number of detections and sample types
Acetaminophen	Pharmaceutical	0
Atrazine	**Herbicide**	4 POCIS
Caffeine	**Pharmaceutical**	4 POCIS
Carbamazepine	**Pharmaceutical**	4 POCIS
Carisoprodol	Pharmaceutical	0
DEET	**Insecticide**	8 POCIS, 2 GW, 1 SW
Diazepam	Pharmaceutical	0
Diclofenac	Pharmaceutical	0
Dilantin	Pharmaceutical	0
Erythromycin	Pharmaceutical	0
Fluoxetine	Pharmaceutical	0
Gemfibrozil	**Pharmaceutical**	4 POCIS
Hydrocodone	Pharmaceutical	0
Ibuprofen	**Pharmaceutical**	2 POCIS, 1 SW
Ketoprofen	**Pharmaceutical**	3 POCIS
Mecoprop	**Herbicide**	2 SW
Memprobamate	Pharmaceutical	0
Norfluoxetine	**Pharmaceutical metabolite**	1 POCIS
Oxybenzone	Personal care (sunscreen)	0
Pentoxifylline	Pharmaceutical	0
Propranolol	**Pharmaceutical**	1 SW
Sulfamethoxazole	Pharmaceutical	0
TCEP	**Flame retardant**	3 POCIS, 1 GW
Triclosan	Personal care (antibacterial)	0
Trimethoprim	**Pharmaceutical**	3 POCIS, 1 GW

Summary

Groundwater is a source of nitrogen for the marine waters of the study area, and larger nutrient inputs were associated with a higher degree of urbanization. Fluxes of dissolved inorganic nitrogen in submarine groundwater discharge from the less urbanized Sandy Hook site were as much as three times greater than fluxes from the less urbanized Sandy Hook site despite a potentially large freshwater submarine groundwater discharge (SGD) rate for the less urbanized site. The SGD nutrient flux estimates for the two Liberty Bay sites were comparable to those previously estimated for the Merrimont Dock site in Lynch Cove, Hood Canal, but were substantially less than the SGD nutrient flux estimated at three sites in Skagit Bay that resulted from large SGD rates and high dissolved inorganic nutrient (DIN) concentrations.

The data suggest that local waters are slightly contaminated with pharmaceutical and personal care products (PPCPs), an effect that likely will increase as population and product usage increases. Use of integrative POCIS samplers was an essential tool as contaminants were present at very low levels (nanograms), which is common for PPCPs in general, but particularly so in an area of only low to moderate urbanization intensity.

Acknowledgments

The authors thank the U.S. Geological Survey Coastal and Marine Geology Program for continued support for the multi-disciplinary Coastal Habitats in Puget Sound program, as well as for work on geologic controls on coastal aquifers. Thanks also to Chris Reich, Jason Greenwood, Marci Marot, and Greg Justin for expert assistance in the field, and to the Suquamish Tribe for site access at Sandy Hook.

References Cited

Alvarez, D.A, Petty, J.D., Huckins, J.N., Jones-Lepp, T.L., Getting, D.T., Goddard, J.P., and Manahan, S.E., 2004, Development of a passive, in situ, integrative sampler for hydrophilic organic contaminants in aquatic environments: Environmental Toxicology and Chemistry, v. 23, no. 7, p. 1640-1648.

Bokuniewicz, Henry, 1980, Groundwater seepage into Great South Bay, New York: Estuarine and Coastal Marine Science v. 10, no. 4, p. 437-444.

Burnett, W.C., Aggarwal, P.K., Aureli, A., Bokuniewicz, H., Cable, J.E.,Charette, M.A., Kontar, E., Krupa, S., Kulkarni, K.M., Loveless, A., Moore, W.S., Oberdorfer, J.A., Oliveira, J., Ozyurt, N., Povinec, P., Privitera, A.M.G., Rajar, R., Ramessur, R.T., Scholten, J., Stieglitz, T., Taniguchi, M., and Turner, J.V., 2006, Quantifying submarine groundwater discharge in the coastal zone via multiple methods: Science of the Total Environment, v. 367, nos. 2–3, p. 498-543.

Burnett, W.C., Cable, J.E., and Corbett, D.R., 2003, Radon tracing of submarine groundwater discharge in coastal environments, in Taniguchi, Makoto, Wang, Kelin, and Gamo, Toshitaka, eds., Land and marine hydrogeology: New York, Elsevier Publications, p. 25-43.

Burnett, W.C., Chanton, J., Christoff, J., Kontar, E., Krupa, S., Lambert, M., Moore, W., O'Rourke, D., Paulsen, R., Smith, C., Smith, L., and Taniguchi, M., 2002, Assessing methodologies for measuring groundwater discharge to the ocean: Eos, Transactions, American Geophysical Union, v. 83, p. 117-123.

Burnett, W.C. and Dulaiova, H., 2003, Estimating the dynamics of groundwater input into the coastal zone via continuous radon-222 measurements: Journal of Environmental Radioactivity v. 69, p. 21-35.

Burnett, W.C., Kim, G., and Lane-Smith, D., 2001, A continuous radon monitor for assessment of radon in coastal ocean waters: Journal of Radioanalytical and Nuclear Chemistry, v. 249, p. 167-172.

Caponne, D.G., and Bautista, M.F., 1985, A groundwater source of nitrate in the nearshore marine sediments: Nature, v. 313, p. 214-216.

Charette, M.A., and Sholkovitz, E.R., 2006, Trace element cycling in a subterranean estuary, Part 2, Geochemistry of the pore water: Geochimica et Cosmochimica Acta, v., 70, p. 811-826.

Conn, K.E., Barber, L.B., Brown, G.K., and Siegrist, R.L., 2006, Occurrence and fate of organic contaminants during onsite wastewater treatment: Environmental Science and Technology, v. 40, no. 23, p. 7358-7366.

Dougherty, J.A., Swarzenski, P.W., Dinicola, R.S., and Reinhard, M., 2010, Occurrence of herbicides and pharmaceutical and personal care products in surface water and groundwater around Liberty Bay, Puget Sound, Washington: Journal of Environmental Quality, v. 39, p. 1173-1180

Dulaiova, H., Burnett, W.C., Chanton, J.P., Moore, W.S., Bokuniewicz, H.J., Charette, M.A., and Sholkovitz, E., 2006, Assessment of groundwater discharges into West Neck Bay, New York, via natural tracers: Continental Shelf Research, v. 26, p. 1971-1983.

Dulaiova, H., Peterson, R., Burnett, W.C., and Lane-Smith, D., 2005, A multidetector continuous monitor for assessment of 222Rn in the coastal ocean: Journal of Radioanalytical and Nuclear Chemistry, v. 263, no. 2, p. 361-365.

Geary, P., 2005, Effluent tracing and the transport of contaminants from a domestic septic system: Water Science and Technology 2005, v. 51, no. 10, p. 283-290.

Glassmeyer, S.T., Furlong, E.T., Kolpin, D.W., Cahill, J.D., Zaugg, S.D., Werner, S.L., Meyer, M.T., and Kryak, D.D., 2005, Transport of chemical and microbial compounds from known wastewater discharges—Potential for use as indicators of human fecal contamination: Environmental Science and Technology, v. 39, no. 14, p. 5157-5169.

Godfrey, E., Woessner, W.W., and Benotti, M.J., 2007, Pharmaceuticals in on-site sewage effluent and ground water, western Montana: Ground Water, v. 45, no. 3, p. 263-271.

Hinkle, S.R., Weick, R.J., Johnson, J.M., Cahill, J.D., Smith, S.G., and Rich, B.J., 2005, Organic wastewater compounds, pharmaceuticals, and coliphage in ground water receiving discharge from onsite wastewater treatment systems near La Pine, Oregon—Occurrence and implications for transport: U.S. Geological Survey Scientific Investigations Report 2005-5055, 98 p.

Kolpin, D.W., Furlong, E.T., Meyer, M.T., Thurman, E.M., Zaugg, S.D., Barber, L.B., and Buxton, H.T., 2002, Pharmaceuticals, hormones, and other organic wastewater contaminants in U.S. streams, 1999–2000—A National Reconnaissance: Environmental Science and Technology 2002, v. 36, no. 6, p. 1202-1211.

Lapointe, B.E, O'Connell, J.D., and Garrett, G.S., 1999, Nutrient couplings between on-site sewage disposal systems, groundwaters, and nearshore surface waters of the Florida Keys: Biodegradation, v. 10, p. 289-307

Manheim, F.T., Krantz, D.E., and Bratton, J.F., 2004, Studying groundwater under DELMARVA coastal bays using electrical resistivity: Ground Water, v. 42, no. 7, p. 1052-1068.

Michael, H.A., Mulligan, A.E., and Harvey, C.F., 2005, Seasonal oscillations in water exchange between aquifers and the coastal ocean: Nature, v. 436, p. 1145-1148. (Also available at http://dx.doi.org/10.1038/nature03935.)

Moore, W.S., 1996, Large groundwater inputs to coastal waters revealed by 226Ra enrichments: Nature, v. 380, p. 612-614. (Also available at http://dx.doi.org/10.1038/380612a0.)

Robinson, C., Gibbes, B., Carey, H., Li, L., 2007a, Salt-freshwater dynamics in a subterranean estuary over a spring-neap tidal cycle: Journal of Geophysical Research, v. 112, C09007, 15 p. (Also available at http://dx.doi.org/10.1029/2006JC003888.)

Robinson, C., Li, L., and Prommer, H., 2007b, Tide-induced recirculation across the aquifer-ocean interface: Water Resources Research, v. 43, W07428, 14 p. (Also available at http://dx.doi.org/10.1029/2006WR005679.)

Swartz, C.H., Reddy, Sharanya, Benotti, M.J., Yin, Haifei, Barber, L.B., Brownawell, B.J., and Rudel, R.A., 2006, Steroid estrogens, nonylphenol ethoxylate metabolites, and other wastewater contaminants in groundwater affected by a residential septic system on Cape Cod, MA: Environmental Science and Technology, v. 40, no. 16, p. 4894-4902. (Also available at http://pubs.acs.org/doi/full/10.1021/es052595%2B.)

Swarzenski, P.W., 2007, U/Th series radionuclides as coastal groundwater tracers: Chemical Reviews, v. 107, no. 2, p. 663-674. (Also available at http://dx.doi.org/10.1021/cr0503761.)

Swarzenski, P.W., and Baskaran, Mark, 2007, Uranium distribution in the coastal waters and pore waters of Tampa Bay, Florida: Marine Chemistry, v. 104, issue 1-2, p. 43-57. (Also available at http://dx.doi.org/10.1016/j.marchem.2006.05.002.)

Swarzenski, P.W., Burnett, W.C., Greenwood, W.J., Herut, B., Peterson, R., Dimova, N., Shalem, Y., Yechieli, Y., and Weinstein, Y., 2006, Combined time-series resistivity and geochemical tracer techniques to examine submarine groundwater discharge at Dor Beach, Israel: Geophysical Research Letters, v. 33, L24405, 6 p. (Also available at http://dx.doi.org/10.1029/2006GL028282.)

Swarzenski, P.W., Kruse, Sarah, Reich, Chris, and Swarzenski, W.V., 2007a, Multi-channel resistivity investigations of the fresh water-saltwater interface—A new tool to study an old problem, in Sanford, W., Langevin, C., Polemio, M., and Povinec, P., eds., A new focus on groundwater–seawater interactions: IAHS Publication, v. 312, p. 100-108.

Swarzenski, P.W., Reich, C., Kroeger, K.D., and Baskaran, M., 2007b, Ra and Rn isotopes as natural tracers of submarine groundwater discharge in Tampa Bay, Florida: Marine Chemistry, v. 104, p. 69-84.

Swarzenski, P.W., Reich, C.D., Spechler, R.M., Kindinger J.L., and Moore, W.S., 2001, Using multiple geochemical tracers to characterize the hydrogeology of the submarine spring off Crescent Beach, Florida: Chemical Geology, v. 179, nos. 1–4, p. 187-202. (Also available at http://dx.doi.org/10.1016/S0009-2541(01)00322-9.)

Swarzenski, P.W., Simonds, F.W., Paulson, A.J., Kruse, S., and Reich C., 2007c, Geochemical and geophysical examination of submarine groundwater discharge and associated nutrient loading estimates into Lynch Cove, Hood Canal, WA: Environmental Science and Technology, v. 41, p. 7022-7029.

Taniguchi, Makoto, 2002, Tidal effects on submarine groundwater discharge into the ocean: Geophysical Research Letters, v. 29, no. 12, p. 1561. (Also available at http://dx.doi.org/10.1029/2002GL014987.)

Taniguchi, Makoto, Burnett, W.C., Smith, C.F., Paulsen, R.J., O'Rourke, Daniel, Krupa, S.L., and Christoff, J.L., 2003, Spatial and temporal distributions of submarine groundwater discharge rates obtained from various types of seepage meters at a site in the Northeastern Gulf of Mexico: Biogeochemistry, v. 66, nos. 1-2, p. 35-53. (Also available at http://dx.doi.org/10.1023/B:BIOG.0000006090.25949.8d.)

Tesoriero, A.J., and Voss, F.D., 1997, Predicting the probability of elevated nitrate concentrations in the Puget Sound Basin—Implications for aquifer susceptibility and vulnerability: Ground Water, v. 35, no. 6, p. 1029-1039. (Also available at http://dx.doi.org/10.1111/j.1745-6584.1997.tb00175.x.)

Vaccaro, J.J., Hansen, A.J., Jr., and Jones, M.A., 1998, Hydrogeologic framework for the Puget Sound aquifer system, Washington and British Columbia: U.S. Geological Survey Professional Paper 1424-D, 77 p., 1 pl. (Also available at http://pubs.er.usgs.gov/publication/pp1424D.)

Vanderford, B.J., Pearson, R.A., Rexing, D.J., and Snyder, S.A., 2003, Analysis of endocrine disruptors, pharmaceuticals, and personal care products in water using liquid chromatography/tandem mass spectrometry: Analytical Chemistry, v. 75, p. 6265-6274.

Weinstein, Y., Burnett, W.C., Swarzenski, P.W., Shalem, Y., Yechieli, Y., and Herut, B., 2007, Role of aquifer heterogeneity in fresh groundwater discharge and seawater recycling—An example from the Carmel coast, Israel: Journal of Geophysical Research, v. 112, C12016, 12 p. (Also available at http://dx.doi.org/10.1029/2007JC004112.)

Suggested Citation

Dinicola, R.S., Swarzenski, P.W., and Dougherty, Jennifer, 2011, Aquatic environment—Circulation, water quality, and phytoplankton concentration, chap. 3 of Takesue, R.K., ed., Hydrography of and biogeochemical inputs to Liberty Bay, a small urban embayment in Puget Sound, Washington: U.S. Geological Survey Scientific Investigations Report 2011–5152, p. 41-52.

Chapter 4. Liberty Bay Sediment and Contaminants

By Renee K. Takesue[1] and Richard S. Dinicola[2]

Introduction

Sources of Sediment to Liberty Bay and Point Bolin

The surficial geology of the Puget Lowlands consists of poorly sorted glacial deposits that form bluffs as great as 100 m thick along the coast (Booth, 1994). Local bluff erosion is the primary mechanism that supplies sediment to Puget Sound beaches (Downing, 1983). Following bluff collapse, small transportable sediment particles are entrained into alongshore or cross-shore flows whereas large particles such as cobbles and boulders generally remain at the toe of the bluff. A much smaller volume of sediment that is sand-sized and coarser enters Puget Sound nearshore in the suspended load of rivers and streams (Downing, 1983).

Shoreline armoring, which is intended to stabilize the toes of bluffs and to prevent beach erosion, reduces upland sediment supply to the nearshore by impounding sediment behind man-made structures such as seawalls, bulkheads, and riprap (MacDonald, and others 1994). In Liberty Bay, about one-half of the shoreline is armored with bulkheads or riprap. Several small creeks supply sand and mud to the nearshore region. This fluvial sediment forms extensive mudflats offshore of Dogfish Creek at the head of Liberty Bay and in deltas at the mouths of Bjorgen and Lemolo Creeks (fig. 4-1). Coastal bluffs 10–20 m high form the western, generally unarmored, shore of Point Bolin.

Sediment Properties and Nearshore Habitat

The sediment grain-size distribution of beaches depends on the sediment source and post-depositional reworking. The sediment grain-size parameter is useful to characterize nearshore habitats for several reasons. First, using the grain-size parameter provides information about the strength of waves and currents (Davis, 1992). Waves and currents remove small particles (sand and mud) more easily than large particles (gravel); thus, a beach composed of large grains has experienced stronger waves and currents than a beach composed of small grains. Second, the sediment grain-size distribution determines the suitability of the substrate as habitat for benthic organisms. Certain organisms such as algae, kelp, and mussels require hard substrate, including cobble- and boulder-sized particles, whereas other organisms such as seagrass, clams, and polychaete worms require unconsolidated, or soft, substrate, such as sand and mud. Third, re-suspension of silt and clay can negatively affect benthic communities. Suspended sediment diminishes underwater irradiance for aquatic plants and can clog the feeding apparatus of filter-feeding organisms. Lastly, the sediment grain-size is related to contaminant data. Fine sediment (silt and clay) can have several orders of magnitude more contaminants associated with it than does sand (Horowitz, 1991). Thus, all other factors being equal, a high percentage of fines can result in high contaminant levels.

Sedimentary Contaminants in Urban Areas

Anthropogenic activities introduce contaminants such as industrial metals and a wide range of organic pollutants—petroleum products (hydrocarbons), pesticides, household chemicals, and pharmaceutical and personal care products—into the environment. Many of these compounds have an affinity for particles rather than water, so when contaminant-bearing runoff from land enters the nearshore, contaminants adsorb onto sediment. Consequently, nearshore sediment is a sink for many contaminants. At greater levels, metals and organic pollutants may pose a health risk to nearshore biota and to humans.

Liberty Bay has had a history of sewage inputs from aging pipelines (Branaman, 2005) and from malfunctioning septic systems (Kitsap County, 2006). Four of seven small creeks flowing into Liberty Bay failed to meet Kitsap County Health District standards for fecal coliform bacteria and dissolved oxygen (Kitsap County, 2006).

[1] U.S. Geological Survey, USGS Pacific Coastal and Marine Science Center, 400 Natural Bridges Drive, Santa Cruz, CA 95060.

[2] U.S. Geological Survey, Washington Water Science Center, 934 Broadway, Tacoma, WA 98402.

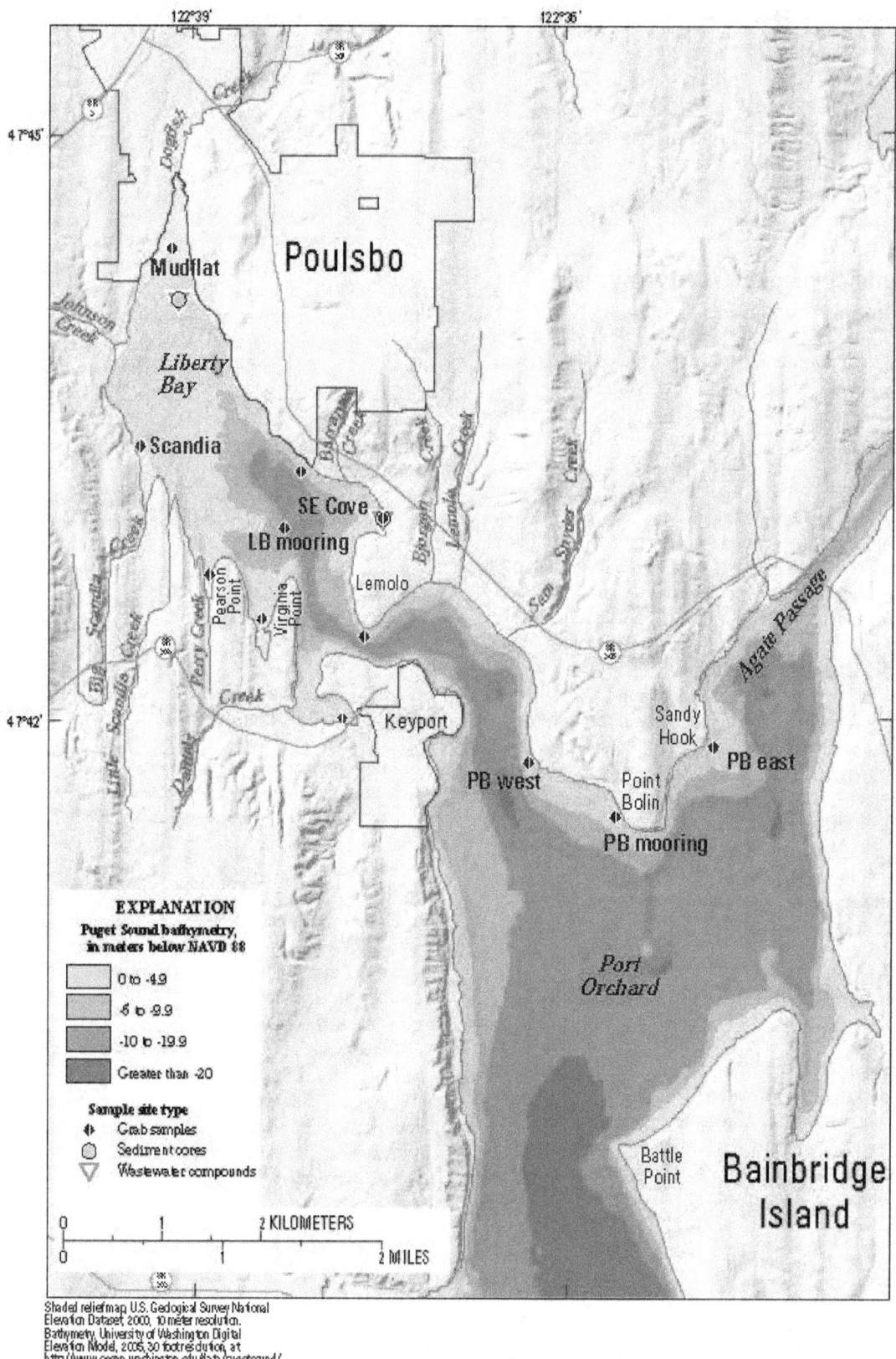

Figure 4-1.   Study area and sediment collections sites, Liberty Bay and vicinity, Central Puget Sound, Washington.

Purpose and Scope

The two goals of this study were to determine grain-size distributions on beaches around Liberty Bay and Point Bolin and to determine whether metals and organic contaminants in Liberty Bay sediment were related to land-use in coastal watersheds, particularly commercial activities in the City of Poulsbo and domestic wastewater leakage into Liberty Bay.

Methods

Sediment Collection

A hand-held clam-shell-type grab sampler was used to collect sediment from eight nearshore sites inside Liberty Bay, from the instrument mooring sites in the center of Liberty Bay (station LB) and on the western side of Point Bolin (station PB), and from an eelgrass bed on the eastern side of Point Bolin at Sandy Hook (fig. 4-1). Grab samples were collected when the tidal height was between +1 and +3 m mean lower-lower water (MLLW) in April 2006. Sediment grain-size distributions and sedimentary trace metals were determined in all grab samples. Two additional sediment samples were collected inside Liberty Bay for determination of sedimentary wastewater indicator compounds: one offshore of Dogfish Creek and one in the southeastern cove between Barrantes Creek and Lemolo peninsula (fig. 4-1). At low tide, short push-cores (15-cm long, 7.6-cm diameter) were collected from a mudflat offshore of Dogfish Creek at the head of Liberty Bay and at a reference site on the eastern side of Point Bolin. The eastern side of Point Bolin was selected as a reference site for sedimentary contaminants because it does not receive sediment from Liberty Bay because of the divergence of littoral drift cells at the tip of Point Bolin (Washington State Department of Ecology, 1991).

Sediment Grain Size Determinations

Sediment grain-size distributions were determined by dry sieving. Approximately 3 g of sediment were dried overnight at 105 °C, disaggregated in a mortar and pestle, and shaken through stainless steel sieves for 12 minutes into greater than 500 μm (coarse sand), greater than 250 μm (medium sand), greater than 150 μm (fine sand), greater than 63 μm (very fine sand), and less than 63 μm (silt and clay, "fines") size fractions. Each size fraction was weighed separately and reported as percentage of total dry weight (percentage weight). The fine fraction was reserved for elemental analyses.

Sedimentary Trace Elements

Total sedimentary major, minor, and trace element contents were measured in the fine fraction (less than 63 μm) to preclude artifacts because of differences in sand content among sites. About 15 mg of dry sediment were digested in a 10 mL mixture of 1:3 hydrofluoric acid and nitric acid in Teflon® vessels using a microwave digestion unit (CEM Corporation). Acid-digestion solutions were evaporated to dryness at 180 °C, and residues were reconstituted with 2 percent nitric acid containing a 100 parts per billion germanium internal standard. Sediment solutions were analyzed by inductively coupled plasma mass spectrometry (ICP-MS, Thermo Finnigan Element) and normalized by the internal standard to account for mass drift. Reference materials consisting of marine sediment (NIST 2702), estuarine sediment (NIST 1646a), and stream sediment (STSD-2, STSD-3) were processed in parallel with unknown samples and were used as external standards to calculate linear relationships between ICP-MS count intensities and elemental contents. One stream sediment sample was used as a consistency standard to estimate procedural and analytical uncertainties.

Sedimentary Wastewater Indicator Compounds

About 500 mL of surface sediment from the mudflat at the head of Liberty Bay and from the southeast cove (fig. 4-1) was placed in pre-certified sample jars (I-CHEM 320 Series) and frozen during storage and transport. The samples were analyzed for sedimentary wastewater indicator compounds at the U.S. Geological Survey National Water Quality Laboratory using methods described by Burkhardt and others (2006).

Results

Grain-Size Distributions

Nearshore sediment was recovered from depths between -0.5 m MLLW and +1.0 m MLLW inside Liberty Bay and from approximately -1.0 m MLLW around Point Bolin, except along the eastern shore of Liberty Bay between the yacht club and the Liberty Bay Marina, where the upper beach consisted of consolidated cobble veneer.

The proportion of fines (silt and clay) was higher inside Liberty Bay than around Point Bolin (fig. 4-2, table 4-1). Along the western shore of Liberty Bay the fine fraction increased from north (Scandia, 14 percent) to south (Keyport, 65 percent). The unconsolidated sediment in the area near Barrantes Creek was the coarsest of the 12 sites, with a median grain size in the medium-sand size-fraction (table 4-1). Bottom sediment at station LB was composed of 91 percent fines, the highest amount reported, whereas the proportion of fines was the lowest, 0 percent, in the bottom sediment at station PB.

Sedimentary Metals

Surface Sediment

Two spatial patterns were identified for trace metal contents in bottom sediment. The first pattern was metals associated with industrial materials—chromium (Cr), lead (Pb), manganese (Mn), nickel (Ni), and vanadium (V)—were increased near a wrecked car on the western shore of Point Bolin. The second pattern consisted of greater contents of elements that form sulfide minerals—arsenic (As), cadmium (Cd, fig. 4-3), copper (Cu), iron (Fe), molybdenum (Mo), and zinc (Zn). These elements were enriched at sites with high proportions of fine sediment that smelled strongly of hydrogen sulfide. No surface sediment metal contents exceeded levels of concern defined by Washington State (Washington State Department of Ecology, 1995).

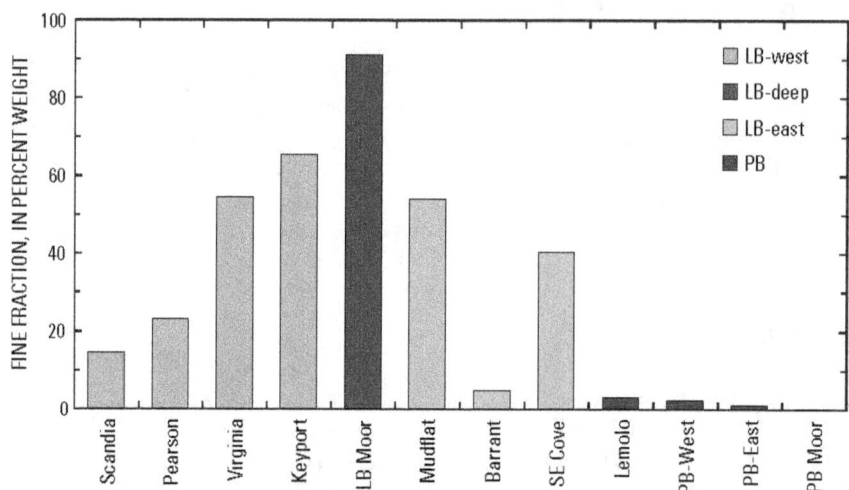

Figure 4-2. Proportion of fines in surface sediment around Liberty Bay (LB) and Point Bolin (PB), Central Puget Sound, Washington.

Table 4-1. Grain size distributions of surface sediment.

[All values reported as percent fractional weight. Values in **bold** indicate the size fraction that corresponds to the median grain size]

Sample site	Coarse sand (CS)	Medium sand (MS)	Fine sand (FS)	Very fine sand (VFS)	Fines (F)	Median grain size (MGS)
Inside Liberty Bay						
Scandia	5	21	**41**	18	14	1.5
Pearson Point	2	9	**43**	23	23	1.9
Virginia Point	0	1	13	32	**55**	3.1
Keyport	0	7	4	24	**65**	3.3
LB Mooring	0	0	1	8	**91**	3.6
Mudflat	0	2	16	29	**54**	3.1
Barrantes Creek	20	**28**	38	8	5	1.0
SE Cove	0	5	19	**36**	40	2.8
Lemolo	8	23	**57**	8	3	1.3
Outside Liberty Bay						
Point Bolin-West	0	6	**55**	36	2	1.8
Point Bolin-East	8	39	**47**	5	1	1.1
Point Bolin Mooring	8	38	**52**	2	0	1.1

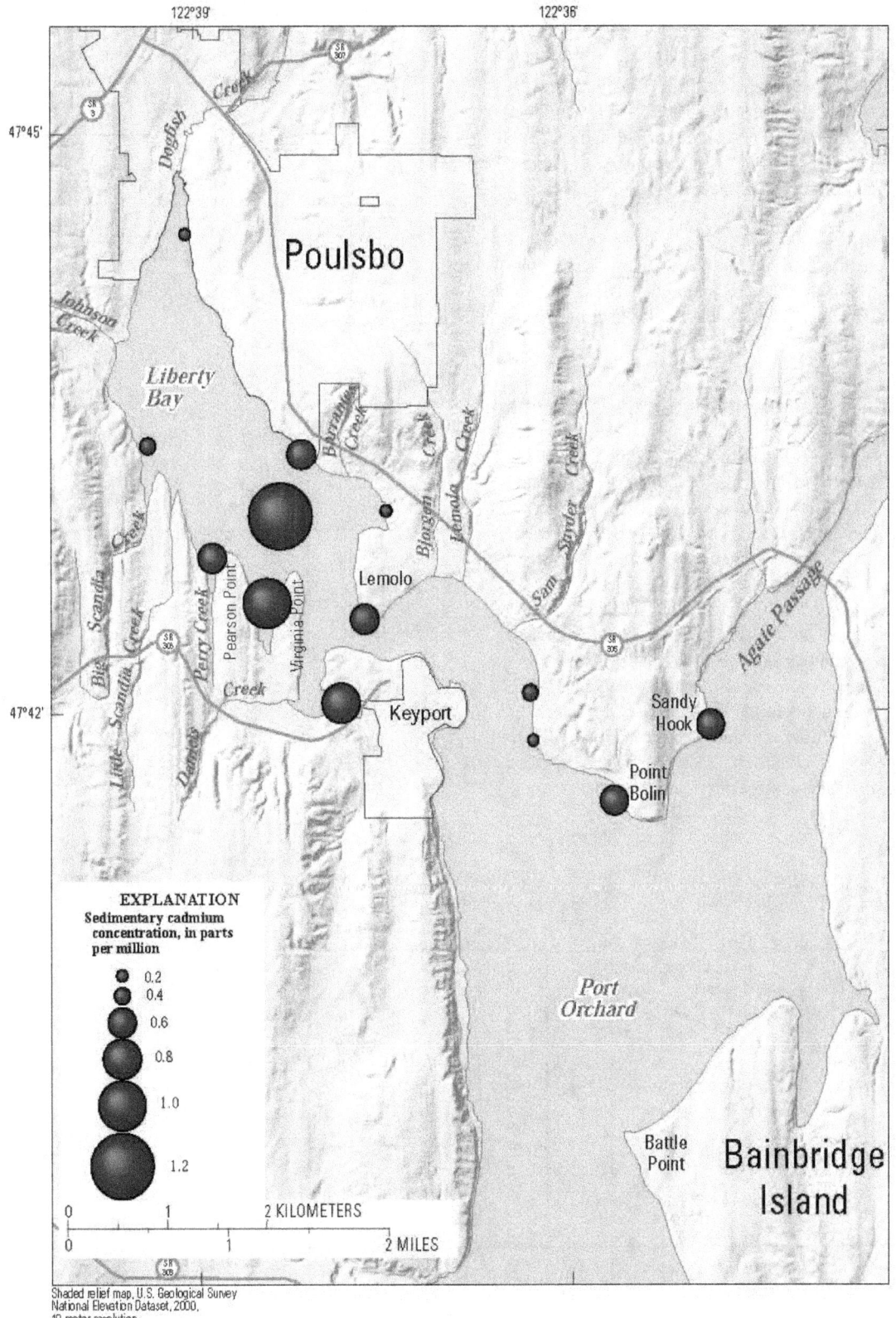

Figure 4-3.   Spatial variation of cadmium (Cd) in surface sediment, Liberty Bay and vicinity, Central Puget Sound, Washington.

Sediment Cores

Contents of Cr, Cu, Ni, Pb, and Zn in two short push-cores were greater at the head of Liberty Bay than at a reference site on the eastern shore of Point Bolin. In Liberty Bay, these metals were highest at the surface and decreased with depth (fig. 4-4, table 4-2). Sedimentary As, Cd, and cobalt (Co) contents were similar at the head of Liberty Bay and at Point Bolin (table 4-2), and showed little variation down-core. The contents of Fe, Mn, and V were higher at Sandy Hook than in Liberty Bay and showed no clear down-core trends (table 4-2). At this time, we do not know whether differences in Fe, Mn, and V contents were the result of anthropogenic or natural factors.

Sedimentary Wastewater Indicator Compounds

Twenty-two wastewater indicator compounds were detected at the head of Liberty Bay and sixteen compounds were detected in the southeast cove (table 4-3). Polycyclic aromatic hydrocarbons (PAHs) were the most commonly detected class of compounds at both sites. No flame-retardants or household wastewater chemicals were detected at either site. Based on the 10-compound criteria of Embrey (2001), sediment near Dogfish Creek was clearly affected by wastewater, and sediment near the mouth of the bay likely was affected by wastewater even though detergent metabolites were not detected. Most wastewater indicator compound concentrations were higher in sediment offshore of Dogfish Creek than in the southeastern part of the bay. Wastewater indicator compounds were not measured in Point Bolin sediment.

Table 4-2.   Trace metals and fine fraction in sediment cores

[All values in parts per million. Even intervals of the Liberty Bay core were analyzed (0–10 cm, n=5). All intervals of the reference core were analyzed (0–11 cm, n=11). **Abbreviation:** SE, standard error].

Trace metals	Liberty Bay		Reference site	
	Mean	SE	Mean	SE
Percent fines	2.7	0.4	1.9	0.3
Chromium (Cr)	896	221	320	71
Copper (Cu)	152	33	41	20
Nickel (Ni)	436	116	81	33
Lead (Pb)	189	53	36	24
Zinc (Zn)	116	10	80	9
Arsenic (As)	5.5	0.5	4.6	0.2
Cadmium (Cd)	0.4	0.0	0.5	0.0
Cobalt (Co)	11.2	0.4	13.0	0.3
Iron (Fe)	34,593	1,483	47,766	1,293
Manganese (Mn)	599	19	1,013	30
Vanadium (V)	106	5	168	5

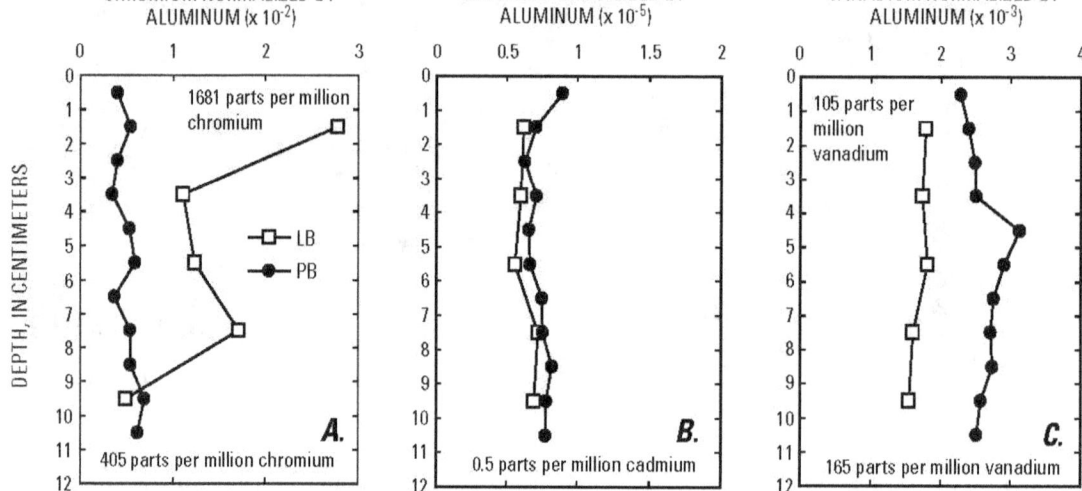

Figure 4-4.   Downcore variations in (*A*) chromium (Cr), (*B*) cadmium (Cd), and (*C*) vanadium (V). Contents were normalized by aluminum (Al), a major constituent of clay, to account for any differences in silt:clay ratios in the fine fraction. Aluminum-normalized metal contents show variations resulting from environmental, rather than geologic, factors.

Table 4-3. Concentrations of wastewater indicator compounds in nearshore sediment, Liberty Bay and vicinity, Central Puget Sound Washington, April 29, 2006.

[All concentrations reported in micrograms per kilogram (µg/kg). Compounds positively detected are indicated in **bold**. **Possible uses or sources:** FR, flame retardant; CP, combustion product. **EDP**, endocrine-disrupting potential; S, suspected; K, known; –, none known. **Compound class codes:** (from Glassmeyer and others, 2005): 3, plant/animal sterol (fecal indicators); 4, detergents and their metabolites; 5, flame retardants; 6, household wastewater; 7, flavors and fragrances; 8, industrial compounds; 9, polycyclic aromatic hydrocarbons (PAHs); 10, pesticides. **Abbreviations:** <, less than; >, greater than; E, estimated value; MRL, minimum reporting level; MDL, minimum detection level]

Compound name	Possible uses or sources	EDP	Compound class code	Concentration		MRL	MDL
				LB6[1]	LB9[2]		
3-beta-coprostanol	**Carnivore fecal indicator**	–	3	**E39.7**	<300	**300**	**216**
beta-sitosterol	**Plant sterol**	–	3	**E467**	**E385**	**300**	**217.8**
beta-stigmastanol	**Plant sterol**	–	3	**E150**	**E115**	**300**	**220.2**
cholesterol	**Often a fecal indicator**	–	3	**E216**	**E402**	**150**	**100.8**
4-cumylphenol	Nonionic detergent metabolite	K	4	<30.0	<30.0	30	20.22
4-n-octylphenol	Nonionic detergent metabolite	K	4	<30.0	<30.0	30	22.08
4-*tert*-octylphenol	Nonionic detergent metabolite	K	4	<30.0	<30.0	30	13.74
4-nonylphenol diethoxylate (sum of all isomers)	**Nonionic detergent metabolite**	K	4	**E245**	<600	**600**	**511.2**
4-nonylphenol monoethoxylate (sum of all isomers)	**Nonionic detergent metabolite**	–	4	**E112**	<300	**300**	**201.6**
4-tert- octylphenol, diethoxylate	Nonionic detergent metabolite	K	4	<30.0	<30.0	30	23.1
4-tert- octylphenol, monoethoxylate	Nonionic detergent metabolite	K	4	<150	<150	150	131.4
para-nonylphenol	**Nonionic detergent metabolite**	K	4	**E118**	<450	**450**	**298.8**
2,2,4,4- tetrabromodiphenyl ether (PBDE 47)	FR	–	5	<30.0	<30.0	30	11.46
tris(2-butoxyethyl) phosphate	FR	–	5	<90.0	<90.0	90	59.1
tris(2-chloroethyl) phosphate	Plasticizer, FR	S	5	<60.0	<60.0	60	42.18
tris(dichloroisopropyl) phosphate	FR	S	5	<60.0	<60.0	60	43.8
tributyl phosphate	Antifoaming agent, FR	S	5	<30.0	<30.0	30	23.58
1,4-dichlorobenzene	Moth repellant; fumigant; deodorant in lavatories	S	6	<30.0	<30.0	30	16.56
3-*tert*-butyl-4-hydroxyanisole (BHA)	Antioxidant, general preservative	K	6	<90.0	<90.0	90	60.6
d-limonene	Fungicide; fragrance	–	6	<30.0	<30.0	30	14.22
n,n-diethyl-meta-toluamide (DEET)	Insecticide; urban uses	–	6	<60.0	<60.0	60	33.72
triclosan	Disinfectant, antimicrobial	S	6	<30.0	<30.0	30	29.76
3-methyl-1h-indole	**Fragrance; stench if feces and coal tar**	–	7	**E13.3**	**E8.69**	**30**	**18.54**
acetophenone	**Fragrance (detergent/tobacco); beverage flavor**	–	7	**E21.5**	<90.0	**90**	**60**
acetyl-hexamethyl-tetrahydronaphthalene	Musk fragrance (widespread), persistent in GW	–	7	<30.0	<30.0	30	7.5
benzophenone	Fixative for perfumes and soaps	S	7	<30.0	<30.0	30	19.08
camphor	Flavor, odorant, ointments	–	7	<30.0	<30.0	30	16.2
hexahydrohexamethyl-cyclo-pentabenzopyran	Musk fragrance (widespread), persistent in GW	–	7	<30.0	<30.0	30	9.9
indole	**Pesticide inert ingredient; coffee fragrance**	–	7	**151**	**77.2**	**60**	**32.1**
isoborneol	Fragrance in perfumes, disinfectants	–	7	<30.0	<30.0	30	23.58
isoquinoline	Flavors, fragrances	–	7	<60.0	<60.0	60	49.86
menthol	Cigarettes, cough drops, liniment; mouthwash	–	7	<30.0	<30.0	30	25.2
bisphenol a	mfg polycarb resins; antioxidant, FR	K	8	U-DELETED	U-DELETED	30	18.72
carbazole	Insecticide; mfg dyes, explosives, and lubricants	–	8	<30.0	<30.0	30	13.44

Table 4-3. Concentrations of wastewater indicator compounds in nearshore sediment, Liberty Bay and vicinity, Central Puget Sound Washington, April 29, 2006.—Continued

[All concentrations reported in micrograms per kilogram (µg/kg). Compounds positively detected are indicated in **bold**. **Possible uses or sources:** FR, flame retardant; CP, combustion product. EDP, endocrine-disrupting potential; S, suspected; K, known; –, none known. **Compound class codes:** (from Glassmeyer and others, 2005): 3, plant/animal sterol (fecal indicators); 4, detergents and their metabolites; 5, flame retardants; 6, household wastewater; 7, flavors and fragrances; 8, industrial compounds; 9, polycyclic aromatic hydrocarbons (PAHs); 10, pesticides. **Abbreviations:** <, less than; >, greater than; E, estimated value; MRL, minimum reporting level; MDL, minimum detection level]

Compound name	Possible uses or sources	EDP	Compound class code	Concentration		MRL	MDL
				LB6[1]	LB9[2]		
diethyl phthalate	**Plasticizer for polymers and resins**	–	8	E7.38	E7.95	60	28.02
diethylhexyl phthalate	Plasticizer for polymers and resins, pesticides	–	8	<150	<150	150	82.8
isophorone	**Solvent for lacquer, plastic, oil, silicone, resin**	–	8	E6.04	<30.0	30	26.04
isopropylbenzene (cumene)	Mfg phenol/acetone, fuels, paint thinner	–	8	<60.0	<60.0	60	51.96
para-cresol	**Wood preservative**	S	8	E49.4	E24.1	150	96.6
phenol	**Disinfectant; mfg many products; leachate**	–	8	E241	E27.3	30	22.98
triphenyl phosphate	Plasticizer; resin; wax; finish; roofing paper, FR	–	8	<30.0	<30.0	30	27.6
1-methylnaphthalene	2–5 percent of gasoline, diesel fuel, or crude oil	–	9	<30.0	<30.0	30	16.68
2,6-dimethylnaphthalene	**Present in diesel/kerosene (trace in gasoline)**	–	9	E14.1	E15.8	30	14.88
2-methylnaphthalene	2–5 percent of gasoline, diesel fuel, or crude oil	–	9	<30.0	<30.0	30	16.68
anthracene	**Wood preservative; component of tar, diesel crude oil; CP**	–	9	E9.18	E5.06	30	11.88
benzo[a]pyrene	**Regulated PAH; used in cancer research; CP**	K	9	E7.90	E6.80	30	14.76
fluoranthene	**Coal tar and asphalt (only trace in gas or diesel); CP**	–	9	E45.8	E35.1	30	13.92
naphthalene	**Fumigant; moth repellant; 10 percent component of gasoline**	–	9	E14.1	E10.7	30	14.1
phenanthrene	**Mfg explosive; tar, diesel fuel, crude oil (CP)**	S	9	E27.7	E15.9	30	12.42
pyrene	**Coal tar and asphalt (only trace in gas or diesel); CP**	–	9	E47.2	E29.4	30	12.36
anthraquinone	**Mfg dye/textiles; seed treatment; bird repellant**	–	10	E9.33	E8.10	30	14.58
atrazine	Selective herbicide	–	10	<60.0	<60.0	60	35.34
bromacil	General use herbicide; >80 percent non-crop on grass/brush	–	10	<300	<300	300	152.4
chlorpyrifos	Insecticide pest and termites (domestic use restricted 2001)	K	10	<30.0	<30.0	30	20.16
diazinon	Insecticide ants, flies, >40 percent non-Ag use	K	10	<30.0	<30.0	30	29.16
metolachlor	General use herbicide, indicator of Ag drainage	–	10	<30.0	<30.0	30	22.32
prometon	Herbicide (non-AG); applied prior to blacktop	–	10	<30.0	<30.0	30	26.52

[1] USGS site ID 474428122390701

[2] USGS Site ID 474301122372501

Discussion

Sedimentary Habitat in Liberty Bay

The retention of fine sediment inside Liberty Bay suggests that waves and currents were weaker in Liberty Bay than around Point Bolin, where bottom sediment contained only a few percent or less of fines. The simplest explanation for high proportions of fine sediment in sheltered coves and in the deep central region of Liberty Bay is that weak waves and currents in these areas allow fine sediment to settle out of suspension and accumulate on the seabed. Where waves and currents are stronger, fine sediment is remobilized and scoured away, resulting in coarsening of the sediment grain-size distribution. The area around Barrantes Creek was an exception to a general trend of decreasing grain size with distance from Dogfish Creek. The beach here is composed predominantly of coarse and medium sand. Extensive clam beds (*Saxidomus gigantea*) also are offshore of Barrantes Creek but not at any other site sampled in April 2006. The lower fraction of mud near Barrantes Creek could have made it a more suitable habitat for filter-feeding *S. gigantea* than the muddier sites around Liberty Bay.

Enrichments of sulfide mineral-forming elements in protected coves and the central bathymetric depression in Liberty Bay indicate that sediment was anoxic, or highly reducing (Bonatti and others, 1970; Morford and Emerson, 1999). Under highly reducing conditions, seawater sulfate is reduced to hydrogen sulfide, which precipitates with sulfide-forming elements. Anoxia and sulfide stress contribute to poor habitat quality for benthic organisms.

Inputs of Contaminants to Liberty Bay

Sediment offshore of Dogfish Creek contained wastewater indicator compounds and trace metals at levels that indicate anthropogenic inputs. Contents of both types of contaminants were greater at the head than at the mouth of Liberty Bay suggesting sources within Liberty Bay or its watershed, rather than a marine source. Such sources could include the three marinas in the City of Poulsbo, live-aboard boats, creosote-treated pilings, and commercial activities. The absence of household wastewater compounds suggests that residential land-use was not an important source of contaminants to nearshore sediment. Commercial and industrial sedimentary contaminants (plasticizers, solvents, lubricants) in the southeast cove probably were transported with fine sediment from the head of the bay because the area around the southeastern cove primarily is residential.

Of the metals enriched at the head of the bay, only Cu typically is greater in household wastewater, yet its enrichment in Liberty Bay sediment was only a factor of 3.7 compared to the reference site, whereas the enrichment factors of 5.4 and 5.3 for other metals, Ni and Pb, were higher. Nickel and Pb are associated with industrial and commercial materials—metal-plating, soldering, batteries, lead-based paint, and pigments. Metal enrichments in the most recently deposited sediment rather than in earlier deposited sediment suggests that surficial runoff or leakage from the municipal sewer line was the source of contaminants, rather than legacy contamination from naval operations at Keyport (May and others, 2005) or the ASARCO smelting plant in Tacoma, which shut down in 1985.

Conclusions

Sedimentary habitat around Liberty Bay and Point Bolin ranged from anoxic mud in sheltered coves and in the deepest part of Liberty Bay to clean sands at the exposed shoreline at the tip of Point Bolin. Sediment anoxia and hydrogen sulfide was associated with low-energy environments, making these areas less-favorable habitat for benthic organisms that cannot tolerate low oxygen and high sulfide levels.

Sedimentary wastewater indicator compounds indicate that waste streams from commercial or light industrial activities (but not residences) and runoff from man-made structures (for example, roads, creosote-treated wood pilings) introduced metals and organic contaminants into Liberty Bay, where they became associated with nearshore sediment. The presence of such compounds in the southeastern cove indicates that sediment and contaminants can be transported from their discharge sites to distal locations and in such a manner have more widespread effects. Greater contents of metals near a wrecked car on a beach illustrate the role of marine debris as a pollution point source.

Acknowledgments

The authors thank Robert Franks of the University of California at Santa Cruz Institute of Marine Sciences for help with ICP-MS analyses, the USGS National Water Quality Laboratory for sedimentary wastewater indicator compound determinations, Steve Cox for help setting up the mobile laboratory, Greg Justin for field support, and Kit Conaway and Bob Rosenbauer for suggestions that improved the manuscript.

References Cited

Bonatti, Enrico, Fisher, D.E., Joensuu, Oiva, and Rydell, H.S., 1970, Postdepositional mobility of some transition elements, phosphorus, uranium, and thorium in deep sea sediments: Geochimica et Cosmochimica Acta, v. 35, no. 2, p. 189-201. (Also available at http://dx.doi.org/10.1016/0016-7037(71)90057-3.)

Booth, D.B., 1994, Glaciofluvial infilling and scour of the Puget Lowland, Washington, during ice-sheet glaciation: Geology, v. 22, no. 8, p. 695-698. (Also available at http://geology.gsapubs.org/content/22/8/695.)

Branaman, J., 2005, Sewage Leak Again Spoils Bay: Kitsap Sun, December 12, 2005.

Burkhardt, M.R., Zaugg, S.D., Smith, S.G., and ReVello, R.C., 2006, Determination of wastewater compounds in sediment and soil by pressurized solvent extraction, solid-phase extraction, and capillary-column gas chromatography/mass spectrometry: U.S. Geological Survey Techniques and Methods, book 5, chap. B2, 40 p. (Also available at http://pubs.usgs.gov/tm/2006/tm5b2/.)

Davis, R.A., 1992, Depositional systems—An introduction to sedimentology and stratigrahy (2d. ed.): Englewood Cliffs, N.J., Prentice-Hall, Inc., 604 p.

Downing, J., 1983, The coast of Puget Sound—Its processes and development: Seattle, Wash., University of Washington Press, 126 p.

Embrey, S.S., 2001, Microbiological quality of Puget Sound basin streams and identification of contaminant sources: Journal of the American Water Resources Association, v. 37, no. 2, p. 407-421.

Glassmeyer, S.T., Furlong, E.T., Kolpin, D.W., Cahill, J.D., Zaugg, S.D., Werner, S.L., Meyer, M.T., and Kryak, D.D., 2005, Transport of chemical and microbial compounds from known wastewater discharges—Potential for use as indicators of human fecal contamination: Environmental Science and Technology, v 39, p. 5157-5169.

Horowitz, A.J., 1991, A primer on sediment-trace element chemistry: Chelsea, Mich., Lewis Publishers, 144 p.

Kitsap County, 2006, Liberty Bay/Miller Bay Watershed 2006 Water Quality Monitoring Report: Port Orange, Wash., Kitsap County Health District, 14 p.

MacDonald, K., Simpson, D., Paulson, B., Cox, J., and Gendron, J., 1994, Shoreline armoring effects on physical coastal processes in Puget Sound, Washington: Olympia, Wash., Shorelands and Water Resources Program, Washington Department of Ecology.

May, C.W., Barrantes, K.B., and Barrantes, L.E., 2005, Liberty Bay nearshore habitat evaluation and enhancement project final report: Lemolo Citizens' Club and Liberty Bay Foundation EPA Non-Point Pollution Funds Grant Project #G0100125.

Morford, J.L., and Emerson, S., 1999, The geochemistry of redox sensitive trace metals in sediments: Geochimica et Cosmochimica Acta, v. 63, p. 1735-1750.

Washington State Department of Ecology, 1995, Chapter 173-204, WAC: Olympia, Washington, Washington State Department of Ecology Publication No. 96-252, 28 p.

Washington State Department of Ecology, 1991, Net shore-drift in Washington State: Olympia, Wash., Shorelands and Coastal Zone Management Program, Washington Department of Ecology, last accessed April 28, 2011, at http://www.ecy.wa.gov/services/gis/data/shore/driftcells.htm.

Suggested Citation

Takesue, R.K. and Dinicola, R.S., 2011, Liberty Bay sediment and contaminants, chap. 4 of Takesue, R.K., ed., Hydrography of and biogeochemical inputs to Liberty Bay, a small urban embayment in Puget Sound, Washington: U.S. Geological Survey Scientific Investigations Report 2011–5152, p. 53-62.

Chapter 5. Eelgrass Habitat near Liberty Bay

By Renee K. Takesue[1]

Introduction

Seagrasses are a widespread type of marine flowering plants that grow in nearshore intertidal and subtidal zones. Seagrass beds are ecologically important because they affect physical, biological, and chemical characteristics of nearshore habitat, and they are sensitive to changes in coastal water quality (Stevenson and others, 1993; Koch, 2001; Martinez-Crego and others, 2008). *Zostera marina*, commonly known as eelgrass, is protected by a no-net-loss policy in Washington State where it may be used as spawning habitat by herring, a key prey species for salmon, seabirds, and marine mammals (Bargmann, 1998). Eelgrass forms broad meadows in shallow embayments or narrow fringes on open shorelines (Berry and others, 2003). Anthropogenic activities that increase turbidity, nutrient loading, and physical disturbance at the coast can result in dramatic seagrass decline (Ralph and others, 2006).

Purpose and Scope

The purpose of this study was to locate and characterize eelgrass beds in Liberty Bay and around Point Bolin, and to compare eelgrass characteristics between urbanized and non-urbanized shorelines. Density of shoreline development was used as an indicator of urbanization.

Methods

In late-April 2006, a visual survey was conducted by small boat to locate eelgrass beds along the shorelines of Liberty Bay and Point Bolin. The shoreline was inspected from north of Keyport, around Liberty Bay, and around Point Bolin to Sandy Hook. Shoot densities, longest-leaf lengths (the canopy height), longest-leaf widths 5 cm above the sheath, and segment lengths of the rhizome chain were measured for *Z. marina* in randomly selected 0.25 m² quadrats along 100 m-long shore-parallel transects. Rhizome segments were added sequentially, and reflect the growth history of the plant. Plant density transects were located at -0.4 m elevation relative to mean lower-lower water (MLLW).

Results

Eelgrass Distribution

No intertidal eelgrass beds were found in Liberty Bay north of Keyport and Lemolo peninsulas. An underwater video survey in 2007 confirmed the absence of intertidal and subtidal eelgrass inside Liberty Bay (J. Gaeckle, Washington Department of Natural Resources, oral commun., 2007). A small fringing bed of *Z. marina* is on the western shore of Point Bolin south of Sam Snyder Creek and an extensive meadow at Sandy Hook on the eastern shore of Point Bolin (fig. 5-1). The depth distributions of the fringing eelgrass bed and the eelgrass meadow differed. On the western side of Point Bolin, where the upper beach is narrow and steep, the fringing bed is entirely intertidal from approximately -0.3 m to -0.6 m MLLW. On the eastern side of Point Bolin, where the beach is flat and broad at Sandy Hook, the eelgrass meadow extends from the intertidal to the subtidal zone, with a minimum depth around 0.0 m MLLW and a maximum depth greater than -1.2 m MLLW.

Plant Growth

Eelgrass plants in the fringing bed were extremely small, with leaves only 23 cm long on average (table 5-1). Species identification confirmed that the plants were indeed *Z. marina* rather than the dwarf species *Z. japonica*, which grows in the high-intertidal zone. In comparison, leaf lengths in the Sandy Hook meadow averaged 41 cm long (table 5-1), typical of intertidal eelgrass in Puget Sound (Phillips, 1984). In April, there were no visible overgrowths of epiphytic diatoms on eelgrass leaves; however, it may have been too early in the season for significant epiphyte accumulation.

[1] U.S. Geological Survey, USGS Pacific Coastal and Marine Science Center, 400 Natural Bridges Drive, Santa Cruz, CA 95060.

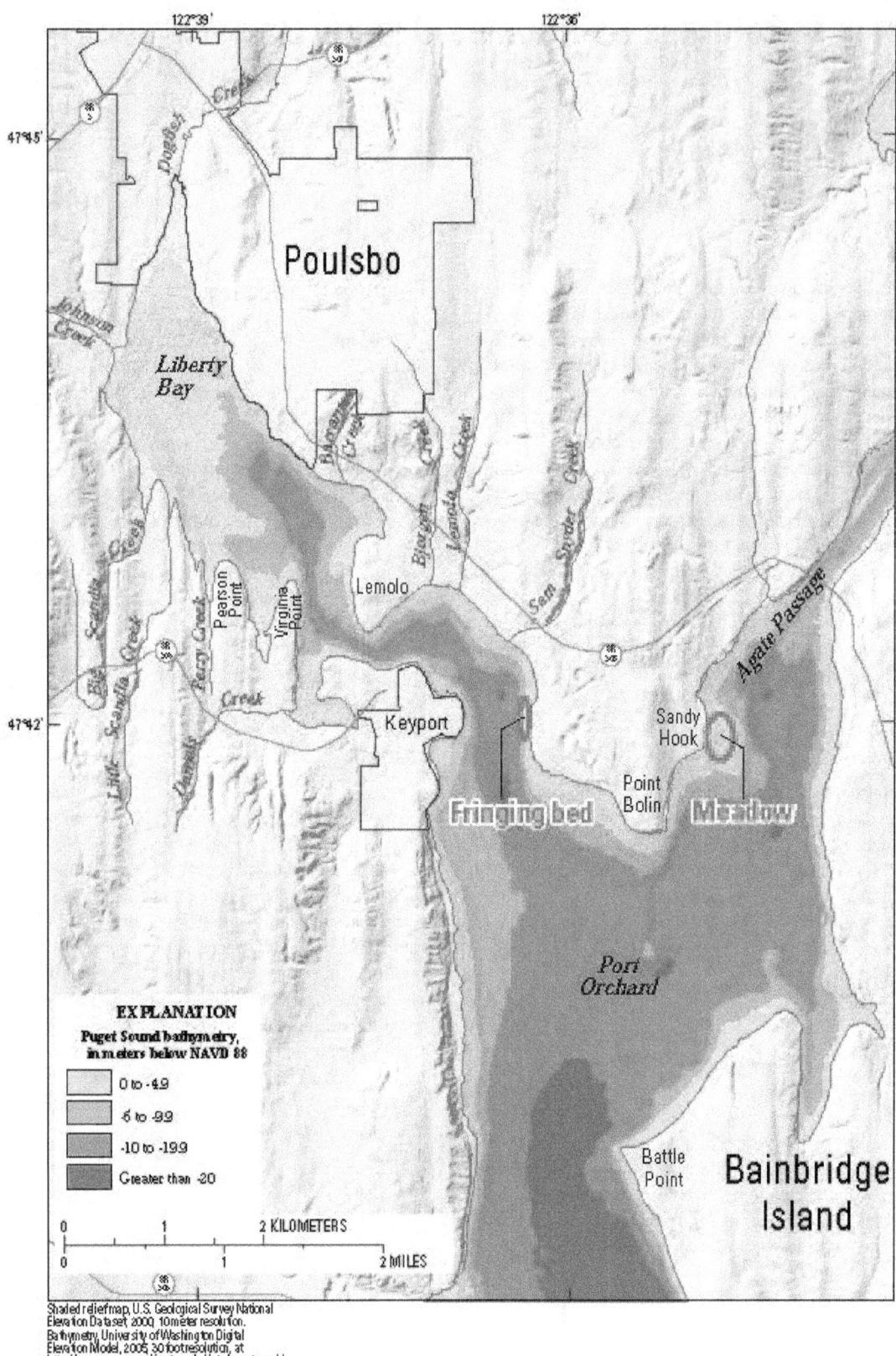

Figure 5-1.   Locations of *Zostera marina* beds around Point Bolin, Central Puget Sound, Washington.

Average growth patterns of eelgrass rhizome chains were similar in the fringing bed and the Sandy Hook meadow (fig. 5-2). After a long period of somewhat constant segment lengths (segments 15-5), lengths of the four most-recently formed segments increased (segments 1–4). Nodes, which intercalate rhizome segments, are more closely spaced in winter and more widely spaced in spring (Phillips, 1984; Olesen and Sand-Jensen, 1994) and form on average every 2–3 weeks in Puget Sound (Phillips, 1984). If segments 1–4 formed at 2-week intervals, then longer rhizome segments grew in March and April, consistent with the expectation of more rapid growth in spring. Absolute elongation rates were twice as high in plants in the Sandy Hook meadow as in plants in the fringing bed. Despite differences in sizes and growth rates of plants, shoot densities were not statistically different at the two sites.

Table 5-1. Eelgrass characteristics in the fringing bed and the Sandy Hook meadow in April 2006.

[**Abbreviations:** MLLW, mean lower low water; SE, standard error; m, meter; cm, centimeter; mm, millimeter; m^2, square meter; n, number; <, less than]

Site	Depth (m, MLLW)		Canopy height (cm)		Leaf width (mm)		Shoot density (number per 0.25 m^2)	
	Minimum	**Maximum**	**Mean**	**SE**	**Mean**	**SE**	**Mean**	**SE**
Fringing, n=10	-0.3	-0.6	23	2	3.0	0.2	145	13
Meadow, n=4	0.0	< -1.2	41	4	5.4	0.3	129	20

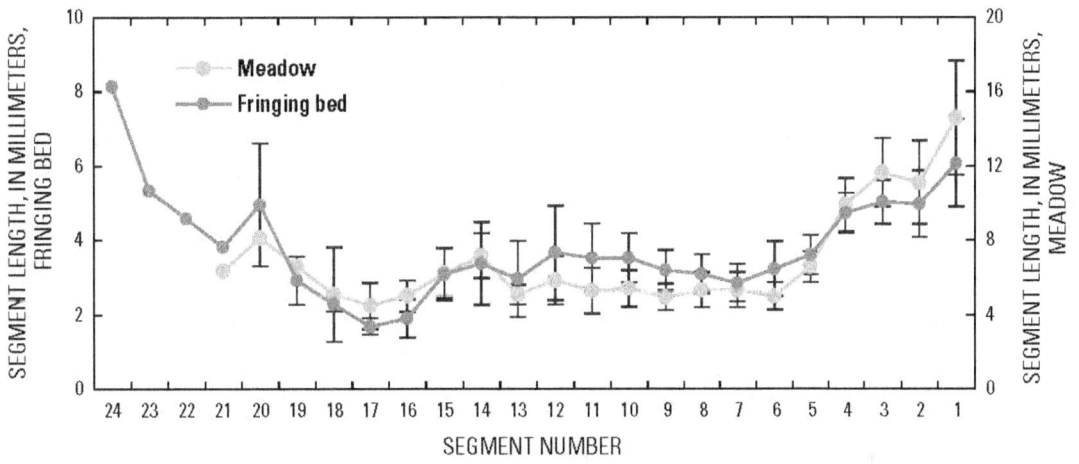

Figure 5-2. Rhizome growth patterns in the fringing bed and the Sandy Hook meadow, Central Puget Sound, Washington. Segment 1 is the most recently formed segment. Note the difference in vertical scales. Error bars show ±1 standard error. Fringing bed (n=10), meadow (n=8).

Discussion

The Importance of Seasonal Light to *Zostera marina* Growth

The synchrony of rhizome biomass growth in both eelgrass beds indicates that a regional factor was the primary control on plant growth. The spring timing of the growth increase suggests that this factor is the seasonal increase in solar irradiance (Olesen and Sand-Jensen, 1994). Some other site-specific factor must account for the size difference between eelgrass plants at the two sites.

Turbidity Limitation on *Zostera marina* Growth

The maximum depth at which *Z. marina* can grow is controlled by underwater light availability (Dennison, 1987; Nielsen and others, 2002). Greater water clarity allows light transmission to greater depths resulting in higher eelgrass abundance at depth compared to sites with lower water clarity (Krause-Jensen and others, 2003). The lower maximum depth of the fringing bed compared to the Sandy Hook eelgrass meadow suggests that below -0.6 m at the fringing site there was insufficient light to support eelgrass growth. The most likely cause of decreased underwater irradiance is high water column turbidity (Aumack and others, 2007). Two creeks (Bjorgen and Lemolo) flow into the nearshore north of the fringing eelgrass bed and were a source of turbid water in April 2006. If this water is advected southward along the shore, it could decrease water clarity over the fringing bed on the western side of Point Bolin.

High turbidity on the western side of Point Bolin also might account for smaller plants in the fringing bed than in the Sandy Hook meadow. Less available light results in less net photosynthesis and biomass growth (Dennison, 1987).

Other Factors that Affect *Zostera marina* Growth

Besides light and turbidity, many environmental factors can limit eelgrass growth (Koch, 2001). These environmental factors include desiccation (Boese and others, 2005), currents in excess of 35 cm/s (Phillips, 1984), wave exposure (Fonesca and Bell, 1998), sediment burial and erosion (Cabaço and others, 2008), prolonged exceedance of optimal ranges of temperature (10–20°C) and salinity (10–25) (Nejrup and Pedersen, 2008), epiphytic algal overgrowths (Short and others, 1995), and pore water sulfide contents greater than 12.8 parts per million (Goodman and others, 1995).

Temperature, salinity, sediment burial, and sediment erosion were excluded as limiting factors on eelgrass growth in Liberty Bay and around Point Bolin. In the absence of large river inflows (low salinity, large sediment flux) or wastewater treatment plant outfalls (low salinity, high temperature) in Liberty Bay or around Point Bolin, water temperature and salinity should not have exceeded optimal ranges for eelgrass growth for long periods.

Around Point Bolin, eelgrass plants did not show signs of burial, which could occur following bluff failure, or sediment erosion, which exposes subterranean rhizomes.

Less desiccation, slower currents, and smaller waves may explain why the minimum depth of the Sandy Hook meadow was shallower than the minimum depth of the fringing bed. Plants at the shallow edge of the Sandy Hook meadow probably were better able to withstand sub-aerial exposure than those in the fringing bed because their leaves were longer and formed an overlapping mat when the tide was out. The leaf mat traps seawater during ebb tide and prevents complete desiccation (Phillips, 1984). Currents and waves probably were attenuated to a greater degree over the Sandy Hook meadow because the dimensions of the meadow were larger than the dimensions of the fringing bed, so plants at the shallow edge of the meadow probably experienced less physical disturbance (Peterson and others, 2004).

At this time, it is not certain how epiphytic algal overgrowths and pore water sulfides may have affected eelgrass growth in the fringing bed and in the Sandy Hook meadow. These factors have a greater effect on eelgrass growth in summer and fall rather than in spring during our study period. Genetic differences also could account for smaller plants in the fringing bed compared to the Sandy Hook meadow (Hämmerli and Reusch, 2002).

Conclusions

From the seasonal nature of rhizome elongation patterns and the regional similarity of these patterns in two eelgrass beds, we conclude that eelgrass biomass growth around Point Bolin primarily was affected by solar irradiance. Minimum and maximum depths of the eelgrass beds and leaf morphology differed widely in the two beds, however, suggesting that site-specific conditions gave rise to these characteristics. Characteristics of eelgrass plants in the fringing bed on the western side of Point Bolin are consistent with a stress response to sub-optimal light levels, possibly arising from higher turbidity. It is not known, however, whether the genotypes for plants in the two beds were different, which also can affect plant morphology and the ability to withstand desiccation and temperature stress at higher tidal elevations. The absence of eelgrass beds inside Liberty Bay precluded a comparison between urban and non-urban shorelines. However, the absence of eelgrass inside Liberty Bay, despite available soft-bottom habitat, suggests that the water column is too turbid to support *Z. marina*.

Acknowledgments

The author thanks Michael Hannam for help conducting eelgrass surveys and plant measurements, Greg Justin for field support, and Kit Conaway and Bob Rosenbauer for suggestions that improved the manuscript.

References Cited

Aumack, C.F., Dunton, K.H., Burd, A.B., Funk, D.W., and Maffione, R.A., 2007, Linking light attenuation and suspended sediment loading to benthic productivity within an Arctic kelp-bed community: Journal of Phycology, v. 43, p. 853-863.

Bargmann, G., 1998, Forage fish management plan—A plan for managing the forage fish resources and fisheries of Washington: Olympia, Wash., Washington Department of Fish and Wildlife, 67 p.

Berry, H., Sewell, A.T., Wyllie-Echeverria, S., Reeves, B.R., Mumford, T.F., Skalski, J.R., Zimmerman, R.C., and Archer, J., 2003, Puget Sound Submerged Vegetation Monitoring Project—2000-2002 monitoring report: Olympia, Wash., Nearshore Habitat Program, Washington State Department of Natural Resources, 60 p. plus appendices.

Boese, B., Robbins, B.D., and Thursby, G., 2005, Desiccation is a limiting factor for eelgrass (*Zostera marina* L.) distribution in the intertidal zone of a northeastern Pacific (USA) estuary: Botanica Marina, v. 48, p. 274-283.

Cabaço, S., Santos, R., and Duarte, C.M., 2008, The impact of sediment burial and erosion on seagrasses—A review: Estuarine, Coastal and Shelf Science, v. 79, p. 354-366.

Dennison, W.C., 1987, Effects of light on seagrass photosynthesis, growth and depth distribution: Aquatic Botany, v. 27, p. 15-26.

Fonseca, M.S., and Bell, S.S., 1998, Influence of physical setting on seagrass landscapes near Beaufort, North Carolina, USA: Marine Ecology Progress Series, v. 171, p. 109-121.

Goodman, J.L., Moore, K.A., and Dennison, W.C., 1995, Photosynthetic responses of eelgrass (*Zostera marina* L.) to light and sediment sulfide in a shallow barrier island lagoon: Aquatic Botany, v. 50, p. 37-47.

Hämmerli, A., and Reusch, T.B.H., 2002, Local adaptation and transplant dominance in genes of the marine clonal plant *Zostera marina*: Marine Ecology Progress Series, v. 242, p. 111-118.

Koch, E.W., 2001, Beyond light—Physical, geological, and geochemical parameters as possible submersed aquatic vegetation habitat requirements: Estuaries, v. 24, no. 1, p. 1-17.

Krause-Jensen, D., Pedersen, M.F., and Jensen, C., 2003, Regulation of eelgrass (*Zostera marina*) cover along depth gradients in Danish coastal waters: Estuaries, v. 26, no. 4A, p. 866-877.

Martinez-Crego, B., Verges, A., Alcoverro, T., and Romero, J., 2008, Selection of multiple seagrass indicators for environmental biomonitoring: Marine Ecology Progress Series, v. 361, p. 93-109.

Nejrup, L.B., and Pedersen, M.F., 2008, Effects of salinity and water temperature on the ecological performance of *Zostera marina*: Aquatic Botany, v. 88, p. 239-246.

Nielsen, S.L., Sand-Jensen, K., Borum, J., and Geertz-Hansen, O., 2002, Depth colonization of eelgrass (*Zostera marina*) and macroalgae as determined by water transparency in Danish coastal waters: Estuaries, v. 25, no. 5, p. 1025-1032.

Olesen, B., and Sand-Jensen, K., 1994, Demography of shallow eelgrass (*Zostera marina*) populations—Shoot dynamics and biomass development: Journal of Ecology, v. 82, p. 379-390.

Peterson, C.H., Luettich, R.A.J., Micheli, F., and Skilleter, G.A., 2004, Attenuation of water flow inside seagrass canopies of differing structure: Marine Ecology Progress Series, v. 268, p. 81-92.

Phillips, R.C., 1984, The ecology of eelgrass meadows in the Pacific Northwest—A community profile: U.S. Fish and Wildlife Service Report FWS/PBS-84/24, 85 p.

Ralph, P.J., Tomasko, D.A., Moore, K.A., Seddon, S., and Macinnis-Ng, C.M.O., 2006, Human impacts on seagrasses—Eutrophication, sedimentation, and contamination, *in* Larkum, A.W.D., Orth, R.J., and Duarte, C.M., eds., Seagrasses: Biology, Ecology and Conservation: Dordrecht, Springer, p. 567-593.

Short, F.T., Burdick, D.M., and Kaldy, J.E., 1995, Mesocosm experiments quantify the effects of eutrophication on eelgrass, *Zostera marina*: Limnology and Oceanography, v. 40, no. 4, p. 740-749.

Stevenson, J.C., Staver, L.W., and Staver, K.W., 1993, Water quality associated with survival of submersed aquatic vegetation along an estuarine gradient: Estuaries, v. 16, p. 346-361.

Suggested Citation

Takesue, R.K., 2011, Eelgrass habitat near Liberty Bay, chap. 5 *of* Takesue, R.K., ed., Hydrography of and biogeochemical inputs to Liberty Bay, a small urban embayment in Puget Sound, Washington: U.S. Geological Survey Scientific Investigations Report 2011-5152, p. 63-68.

Chapter 6. Stable Isotopes of Nitrogen and Carbon as Tools to Monitor Eutrophication and Trophic Dynamics

By Theresa L. Liedtke[1] , Collin D. Smith[1], and Dennis W. Rondorf[1]

Abstract

The human population in Puget Sound is projected to grow about 20 percent between 2010 and 2030. The continuing rapid growth in population will increase urbanization and may be accompanied by anthropogenic eutrophication of Puget Sound waters. Stable isotopes of nitrogen (N) and carbon (C) can be used to indicate the source, utilization, and destination of organic matter in estuarine environments. We tested the feasibility of using stable isotopes of nitrogen and carbon as long-term monitoring tools of eutrophication and food web changes in Puget Sound. For our proof-of-concept study, we selected Liberty Bay, a water body that experiences nitrogen enrichment due to sewage spills and improperly functioning septic systems. Nearby Point Bolin was used as a less perturbed reference site. In cooperation with the Suquamish Tribe and the Liberty Bay Foundation, we collected 140 biological tissue and sediment samples from Liberty Bay and Point Bolin. We found some enrichment of $\delta^{15}N$ among samples of producers such as macroalgae, suggesting an influence of wastewater. Organisms at higher trophic levels were progressively enriched in ^{15}N; however, the interpretation was increasingly uncertain. The $\delta^{15}N$ levels in fish indicated they were enriched less than 1 part per thousand (‰) in Liberty Bay compared to Point Bolin. We attributed the relatively depleted $\delta^{13}C$ levels observed at lower trophic level organisms in Liberty Bay to terrigenous inputs from the forested watershed. The $\delta^{13}C$ level of fish was depleted compared to samples of macroalgae and eelgrass indicating they were not the source of carbon in fish. We concluded that it was more likely that $\delta^{13}C$ in fish were derived from sediment particulate organic matter and phytoplankton. Although the use of stable isotopes has many appealing aspects for long-term monitoring, the relatively wide range in values in the literature indicates there are site-specific, spatial, and temporal sources of variation. We surmise that by carefully selecting primary producers or low-level consumers such as mussels, stable nitrogen and carbon isotopes may be promising candidates for long-term monitoring tools.

Introduction

Increasing urbanization in Puget Sound can have numerous effects on freshwater and marine ecosystems. Watersheds in Puget Sound vary widely in degree of urbanization and density of human population. In studies of a diverse group of watersheds, Cole and others (1993) reported a strong relationship between the density of the human population in a watershed (people/km^2) and the nitrogen loading in rivers draining those watersheds. Many aquatic ecosystems in the Pacific Northwest are nutrient limited and urbanization can alter the structure of the aquatic community, particularly with the addition of various forms of nitrogen. To study the effects of urbanization, we identified the need for a long-term monitoring tool.

Urbanization, like other changes in land use, can have a cascading effect on ecological processes related to biological diversity, food chain length, sustainability, and bioaccumulation of contaminants. The study of trophic relations in aquatic ecosystems is useful to understand a variety of stressors (for example acidity, wastewater, and contaminants), which can decrease species richness and lead to shortened food chains (Woodwell, 1983; Odum, 1985; Ford, 1989). Traditional approaches to evaluations of trophic dynamics were considered and rejected because they likely would be costly and perhaps problematic where tidal currents have the potential to commingle nutrients originating from Puget Sound and from nearby watersheds. Therefore, to study relations between nutrient loading and trophic levels an approach was selected using stable isotopes of nitrogen (N) and carbon (C).

Nitrogen-stable isotopes in plants, macroalgae, particulate organic matter (POM), fish, mussels, and sediment samples can be used as indicators of anthropogenic eutrophication (Lake and others, 2001; Cole and others, 2004). Differences in ratios of ^{15}N to ^{14}N have been used to define food webs, and act as natural tracers of nitrogen sources.

[1]U.S. Geological Survey, Western Fisheries Research Center, Columbia River Research Laboratory, 5501-A Cook-Underwood Road, Cook, WA 98605.

The ratio of ^{15}N to ^{14}N is expressed as δ^{15}N in ‰ units (where δ indicates the deviation from a standard, and ‰ indicates parts per thousand; see Methods section of this chapter). The nitrogen isotope ratio has proven to be a useful tool to assess effects of watershed urbanization on coastal water bodies (Cole and others, 2004). Cole and others (2004) reported that δ^{15}N values of macrophytes increased when wastewater as a percentage of total N load increased in ponds and estuaries. The increase in δ^{15}N values occurs because sewage has higher ^{15}N levels than natural sources of N in ponds and estuaries. Stable N isotopes also are useful in tracing organic matter through food webs. Consumers typically show a 2–4 ‰ increase in δ^{15}N relative to their food source, a fractionation caused by kinetic differences between light and heavy isotopes during metabolism (McClelland and others, 1997).

Similar to the approach used with N, the ratio of stable isotopes of C can be used to understand C sources. The ratio of ^{13}C to ^{12}C is expressed as δ^{13}C in ‰ units (where δ indicates the deviation from a standard, and ‰ indicates parts per thousand; see Methods section in this chapter). At the bottom of the food chain, plants preferentially incorporate the lighter isotope ^{12}C rather than the heavier isotope ^{13}C, and as a result, they produce plant tissue with lower δ^{13}C values than the source of the carbon, atmospheric carbon dioxide (CO_2). Terrestrial plants have several pathways of photosynthesis, resulting in plants such as corn and grains having a δ^{13}C value of about -14 ‰. Other plants of particular interest were the coniferous and deciduous trees, which have a δ^{13}C value of about -26 ‰ (Kendall and others, 2001). Other producers of C, including marine and profundal zones of lakes, tend to have highly negative δ^{13}C values, and as a result of this variation in C fixation, the δ^{13}C values at the base of food chains can vary. Unlike N, the C isotope ratio of consumers is similar to that of their food resources; this relation results in δ^{13}C values being conserved up the food chain with a mean isotopic shift of $+0.5 \pm 0.13$ ‰ (mean ± SE) (Vander Zanden and Rasmussen, 1999; McCutchan and others, 2003).

In this study, we proposed to sample a moderately urbanized embayment, Liberty Bay, and a reference location, nearby Point Bolin on Puget Sound, to analyze biological organisms for δ^{15}N and δ^{13}C values. The objectives were (1) to determine if there was evidence of anthropogenic eutrophication in a small embayment with a relatively high tidal exchange, (2) to evaluate the magnitude of isotopic shifts in organisms in Liberty Bay and at Point Bolin for consideration as a long-term monitoring tool, and (3) to interpret the findings with the spatial and temporal information in the literature.

Methods

Study Site

Of 10 embayments in central Puget Sound considered for this study, Liberty Bay was selected. Liberty Bay has been subjected to occasional sewage spills from a pipe near Poulsbo, Washington, and ammonium (NH_4) levels measured in the bay indicate the early warning signs of eutrophication (Mackas and Harrison, 1997). Such point sources may get significant public attention, but their ecological significance is unknown. Urban development is limited in the watersheds of Liberty Bay, and shoreline development varies from the small urban area of Poulsbo to relatively undeveloped areas, with very limited shoreline modifications near Point Bolin. Other sources of N loading in Liberty Bay may come from changes in land use, lawn-fertilizer runoff, and N-enriched groundwater associated with septic system drain fields.

Sample Collection and Processing

Biological samples at various levels of the food web in Liberty Bay and at the reference site, Point Bolin, were collected for analysis of the stable isotope ratios δ^{15}N and δ^{13}C. A paired sampling approach was used, collecting the same species at both sites so that sites could be compared. Field sampling occurred on April 28 and 29 and June 25, 2006. Samples were collected in Liberty Bay at Oyster Plant Park and near the boardwalk area. Samples were placed on dry ice shortly after collection, and subsequently were stored frozen until processed for isotope analysis. Sediment samples were collected to analyze sediment POM. Sediment was collected from the upper 20 cm of the beach material and was acid-fumigated in a desiccator with 12 molar (M) hydrochloric acid (HCl) prior to analysis (Harris and others, 2001). Plankton samples were collected with a plankton net (80 micron mesh size) and no effort was made to separate zooplankton from abundant blooms of phytoplankton. Macroalgae were collected from shallow intertidal waters and holdfasts were not included. Leaf segments were collected from eelgrass, *Zostera marina*, adrift at the sample sites, and leaves were not cleaned of epiphytes or washed in HCl. Muscle tissue of the sand clam *Macoma secta*, dogwelk *Nucella lamellose*, and bay mussel *Mytilus trossulus* were dissected from animals and processed. The gelatinous tissue of the purple sea star *Pisaster ochraceus* was collected by removing the surface of the central disc of the exoskeleton, removing tissue for the sample, and fumigating in a desiccator with 12 M HCl. The samples from fish were collected from the dorsal epaxial muscle mass. To prepare for analysis sample tissues were dried at 60 °C for at least 48 hours and ground with a mortar and pestle. Samples were placed in pre-weighed metal capsules for stable isotope analysis at the University of California at Davis.

Solid materials were analyzed for ^{13}C and ^{15}N isotopes using a PDZ Europa ANCA-GSL elemental analyzer interfaced to a PDZ Europa 20-20 isotope ratio mass spectrometer (Sercon Ltd., Cheshire, UK). Samples were combusted at $1,020\,°C$ in a reactor packed with chromium oxide and silvered colbatous/cobaltic oxide. Following combustion, oxides were removed in a reduction reactor (reduced copper at $650\,°C$) and the helium carrier flowed through a water trap (magnesium perchlorate) and an optional CO_2 trap (for N-only analyses). The delta ratio of ^{15}N to ^{14}N is expressed as $\delta^{15}N\,‰ = [(R_{sample}-R_{reference})/R_{reference}] \times 1,000$, where R is $^{15}N/^{14}N$ in parts per thousand (‰), and the reference is atmospheric N_2 (Peterson and Fry, 1987). A similar ratio is used to describe the relation of stable isotopes delta $^{13}C/^{12}C$, denoted here as $\delta^{13}C$ in parts per thousand. The reference for $\delta^{13}C$ was a standard representing the Cretaceous fossil *Belemnitella americana* from the PeeDee formation in South Carolina. Because the $\delta^{13}C$ value for the reference material is higher than nearly all other carbon materials, the scale for results presented in this study is entirely of negative delta values.

Analysis

Mean $\delta^{15}N$ and $\delta^{13}C$ values and their standard deviations were reported. Mean differences were tested using the Randomization Test and a bootstrapping technique was used to estimate the 95 percent confidence intervals (CI) for the means (Cassell, 2002). This approach is appropriate for the small sample sizes used in this study. A one-tailed test was used for $\delta^{15}N$ to determine if mean values from Liberty Bay were significantly higher ($P = 0.025$) than mean values from Point Bolin because the hypothesis was that enrichment associated with urbanization and land use would be higher at the more urbanized site. A two-tailed Randomization Test and a similar bootstrapping approach were used to test mean $\delta^{13}C$ at the two sample sites because a specific hypothesis was not made about the relation between the sites. Stable isotope values are presented as bivariate plots of $\delta^{15}N$ and $\delta^{13}C$ values, as plots of each isotope ratio by sample site and species, and as tables of means, standard deviations, and differences in values between sites.

Results and Discussion

Sediment

The $\delta^{15}N$ value for sediment at Point Bolin was higher than the values for samples collected in Liberty Bay, but the difference was not statistically significant. The mean $\delta^{15}N$ values were 7.46 ‰ near Oyster Plant Park in Liberty Bay, 8.96 ‰ near the Boardwalk in Liberty Bay (mean for Liberty Bay = 8.21 ‰), and 10.33 ‰ at Point Bolin (tables 6-1 and 6-2; $P > 0.05$). The relatively high $\delta^{15}N$ values at Point Bolin are not easily explained as originating from eutrophication, but may be more a result of fractionation by trophic levels or geochemical changes in different sediments. Our interpretation of $\delta^{15}N$ levels in sediments between our sample sites was limited by unknown sediment composition and deposition rates. Some marine enrichment associated with natural processes in the estuarine environment was expected such as denitrification, thereby increasing $\delta^{15}N$ levels in the estuary sediments (McClelland and Valiela, 1998). Some enrichment was noted at the boardwalk site relative to Oyster Plant Park, but both of these potential estuary sites in Liberty Bay were less enriched than sites at Point Bolin.

Because the creek at the head of Liberty Bay may transport nitrogen sources into the bay, several potential sources from the watershed were considered. First, synthetic fertilizers were considered but were determined an unlikely source of nitrogen. Fertilizers are isotopically light (closer to zero) because they are manufactured from atmospheric nitrogen, resulting in groundwater with $\delta^{15}N$ values at -3 to +3 ‰ (Freyer and Aly, 1974; Heaton, 1987; Bannon and Roman, 2008). Second, the creek at the head of Liberty Bay may transport POM of terrestrial origin into Liberty Bay. However, the $\delta^{15}N$ values for deciduous and coniferous forests generally range from +3 to +7 ‰ (Kendall and others, 2001). Because the $\delta^{15}N$ values for these forests are less than the $\delta^{15}N$ values for sediment POM reported in Liberty Bay, these forests likely are not the source of N in Liberty Bay sediments.

A third potential source of N input into Liberty Bay is wastewater. In a comparison of wastewater-influenced and reference streams, Singer and Battin (2007) reported benthic fine POM with a $\delta^{15}N$ value of -1.3 ‰ for reference streams and POM $\delta^{15}N$ levels of +8 ‰ for wastewater-influenced streams. We compared our findings in Liberty Bay ($\delta^{15}N$ mean = 8.21 ‰, SE = 0.30) with values in a review of 10 studies (Tucker and others, 1999) and determined that Liberty Bay $\delta^{15}N$ values were higher than marine sediment POM values in most studies (mean = 6.8 ‰; range 4.1–8.9 ‰). An increase in $\delta^{15}N$ level of several parts per thousand may reflect substantial changes in N loading. For example, McClelland and Valiela (1998) report that a relatively modest 1.9 ‰ increase in POM $\delta^{15}N$ values in estuaries of Waquoit Bay, Cape Cod, Massachusetts was associated with increases in the percentage of wastewater in the total N load from 4 to 63 percent. Furthermore, in the shallow near-shore area of the Baltic Sea, Voss and others (2005) considered mean $\delta^{15}N$ values of sediment POM at 7.3 ‰ (+ 2.1 ‰) to be characteristic of trace anthropogenic nitrogen. Although the age or depositional rates of sediments in Liberty Bay are not known, eutrophication has been associated with increased POM $\delta^{15}N$ levels in the short-term and long-term records of sediments. For example, increasing human population was associated with a record of increasing $\delta^{15}N$ levels during the last century (1900 to 2000) in Baltic Sea sediments. Analysis of sediment cores indicated that increases in $\delta^{15}N$

Table 6-1. Locations, mean, and standard deviation values for stable isotope ($\delta^{15}N$ and $\delta^{13}C$) samples collected in Liberty Bay and Point Bolin, Central Puget Sound, Washington, 2006.

[Abbreviations: N, nitrogen; C, carbon; ‰, parts per thousand; SD, standard deviation]

Date collected	Common name	Scientific name	Location	Number of samples	Mean $\delta^{15}N$ (‰)	Standard deviation, ^{15}N (‰)	Mean $\delta^{13}C$ (‰)	Standard deviation, $\delta^{13}C$ (‰)
04-28-06	Starry flounder	*Platichthys stellatus*	Liberty Bay	5	15.30	0.13	-13.64	0.54
			Point Bolin	5	14.40	0.67	-13.43	0.43
06-25-06	Pacific herring	*Clupea pallasii*	Liberty Bay	5	13.96	0.36	-15.90	0.17
			Point Bolin	5	13.30	0.58	-17.33	0.39
06-25-06	Shiner perch	*Cymatogaster aggregata*	Liberty Bay	5	14.15	0.29	-14.03	1.80
			Point Bolin	5	13.34	0.21	-14.71	0.24
06-25-06	Pacific staghorn sculpin	*Leptocottus armatus*	Liberty Bay	5	14.89	0.32	-13.22	0.49
			Point Bolin	5	14.25	0.31	-13.84	0.48
04-29-06	Speckled sanddab	*Citharichthys stigmaeus*	Liberty Bay	5	13.11	0.37	-14.30	0.49
			Point Bolin	5	13.66	0.43	-13.87	0.22
04-28-06	Sand clam	*Macoma secta*	Liberty Bay	5	9.93	0.75	-15.85	0.36
			Point Bolin	5	9.78	0.51	-13.94	1.52
04-28-06	Bay mussel	*Mytilus trossulus*	Liberty Bay	5	7.54	0.93	-17.84	0.27
			Point Bolin	5	6.77	1.29	-17.10	0.31
04-28-06	Dogwhelk	*Nucella lamellosa*	Liberty Bay	5	12.55	0.58	-14.20	0.20
			Point Bolin	5	11.12	0.20	-14.48	0.27
04-29-06	Purple sea star	*Pisaster ochraceus*	Liberty Bay	5	11.38	0.53	-15.56	0.48
			Point Bolin	5	12.08	0.64	-14.28	0.48
04-29-06	Sea lettuce	*Ulva spp.*	Liberty Bay	10	8.17	1.20	-12.82	0.29
			Point Bolin	5	6.16	0.59	-11.72	0.32
04-29-06	Eelgrass	*Zostera marina*	Liberty Bay	5	4.54	1.56	-11.39	0.28
			Point Bolin	10	4.98	0.63	-9.20	0.11
04-29-06	Red algae	*Gracilaria pacifica*	Liberty Bay	5	7.81	0.89	-17.37	0.24
			Point Bolin	5	5.24	1.04	-13.50	0.14
04-29-06	Plankton	*Plankton*	Liberty Bay	5	7.76	2.76	-17.25	1.40
			Point Bolin	5	4.44	3.14	-19.33	1.77
04-29-06	Sediment	*Sediment*	Liberty Bay	10	8.21	0.94	-21.95	1.23
			Point Bolin	5	10.33	2.11	-16.52	0.23

Table 6-2. Results of a randomization test of mean δ^{15}N values and 95 percent confidence interval estimated using a bootstrap approach for samples collected in Liberty Bay and at Point Bolin, Central Puget Sound, Washington, 2006.

[**Abbreviations:** CI, confidence interval; ‰, parts per thousand; NA, not applicable]

	Nitrogen analysis results			Bootstrapped confidence intervals (1,000 iterations)	
Sample type	95 percent critical value (‰)	Observed difference (‰)	*P* value	Liberty Bay 95 percent CI about the mean (‰)	Point Bolin 95 percent CI about the mean (‰)
Sediment	1.5160	-2.1140	0.9867	(7.660, 8.802)	(8.476, 11.716)
Plankton	3.1760	3.3160	0.0417	(6.832, 8.956)	(2.070, 6.910)
G. pacifica	1.6800	2.5680	0.0035	(6.996, 8.345)	(4.425, 5.950)
Z. marina	0.9310	-0.4340	0.7726	(3.508, 5.938)	(4.649, 5.371)
Ulva spp.	1.2880	2.0080	0.0022	(7.462, 8.824)	(5.679, 6.588)
P. ochraceus	0.6900	-0.6980	0.9467	(10.876, 11.754)	(11.550, 12.530)
N. lamellosa	0.8840	NA	<0.0001	(12.072, 12.974)	(10.960, 11.254)
M. trossulus	1.1580	0.7660	0.1499	(6.854, 8.304)	(5.734, 7.776)
M. secta	0.6240	0.1400	0.3801	(9.460, 10.614)	(9.380, 10.192)
C. stigmaeus	0.4960	-0.5560	0.9609	(12.838, 13.418)	(13.332, 14.016)
L. armatus	0.6500	0.6340	0.0693	(14.320, 15.476)	(14.020, 14.536)
C. aggregata	0.5180	NA	<0.0001	(13.924, 14.377)	(13.182, 13.501)
C. pallasii	0.6580	0.6580	0.0485	(13.745, 14.288)	(12.866, 13.750)
P. stellatus	0.6420	NA	<0.0001	(15.206, 15.406)	(13.792, 14.784)

levels ranged from 0 to 2.4 ‰ in open waters and 2.9 to 10.0 ‰ in coastal areas (Voss and Struck, 1997; Voss and others, 2000). Although comparisons of sediment POM δ^{15}N values of Liberty Bay and Point Bolin does not support conclusions about N loading associated with the urban center in the bay, comparisons of sediment POM δ^{15}N values to the literature suggest some eutrophication.

The δ^{13}C axis of bivariate plot for stable isotopes in sediments provided additional information on the origin of sediment POM (fig. 6-1). First, some sources of carbon can be ruled out in the sediment POM. At Point Bolin sediment POM δ^{13}C (mean = -16.52 ‰, SD = 0.23) probably did not originate directly from nearby macroalgae, *Ulva* spp. (mean = -11.72 ‰, SD = 0.32) and *Gracilaria pacifica* (mean = -13.50 ‰, SD = 0.14), or eelgrass *Z. marina* (mean = -9.20 ‰, SD = 0.11; table 6-1). The δ^{13}C values in sediment POM at Point Bolin were relatively depleted compared to macroalgae and *Z. marina*. However, based on δ^{13}C values from the literature, large kelp or benthic algae could not be ruled out as sources of C. The δ^{13}C values of kelp forests in

Norwegian waters ranged from -18.9 to -22.3 ‰ and in varied habitats in the Aleutian Islands from -12.6 to -28.0 ‰ with an overall mean of -17.7 ‰ (SD = 2.3) (Duggins and others, 1989; Simenstad and others, 1993; Fredriksen, 2003). The sediment POM δ^{13}C values at Point Bolin were near values measured for benthic algae (-17 ‰, Petersen and Fry, 1987), microphytobenthos (-16.0 ‰, Sauriau and Kang, 2000), and benthic microalgae (-17 to -15 ‰, Page and Lastra, 2003). The δ^{13}C values of sediment POM samples collected in Liberty Bay (mean = -23.12 and -20.79 ‰) were not as depleted as might be expected from debris from deciduous and coniferous trees (δ^{13}C -28.9 and -26 ‰; Simenstad and Wissmar, 1985; Kendall and others, 2001). However, the δ^{13}C values of sediment POM were similar to δ^{13}C values reported by Simenstad and Wissmar (1985) in estuarine (-23.8 ‰) and marine littoral sediments (-20.3 ‰). The differences between Liberty Bay and Point Bolin δ^{13}C values may be attributed to a mixture of terrestrial inputs into Liberty Bay resulting in depleted δ^{13}C values.

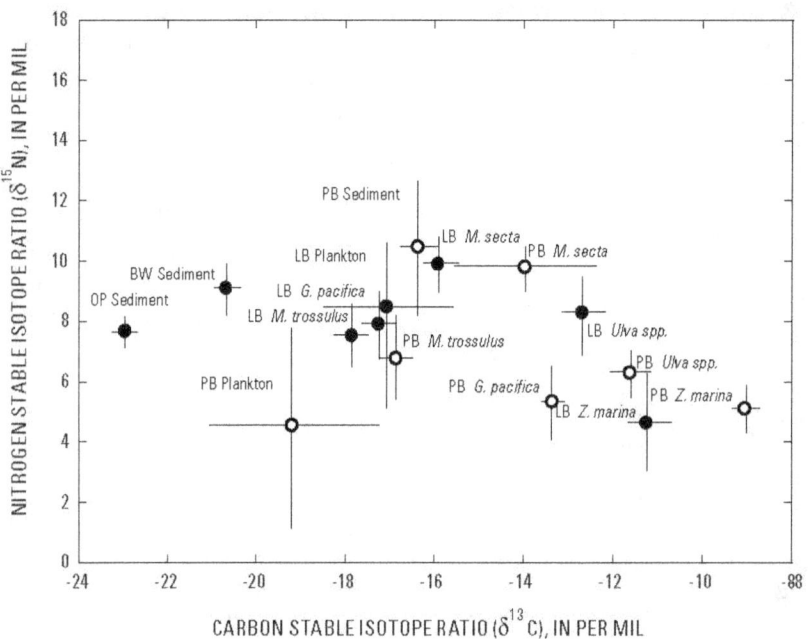

Figure 6-1.   Comparison of mean (±SD) δ^{15}N and δ^{13}C values for algae, plankton, sediment particulate organic matter (POM), mussels, and clam samples collected in Liberty Bay and at Point Bolin, Central Puget Sound, Washington. Sediment samples collected near Oyster Plant Park (OP) and the Boardwalk (BW) were collected in Liberty Bay. See table 6-1 for common names and scientific names.

Eelgrass

The δ^{15}N values for eelgrass *Z. marina*, sea lettuce *Ulva* spp., and red algae *G. pacifica* samples provide a contrast between macroalgae and vascular plants. The δ^{15}N values of *Z. marina* in Liberty Bay (mean = 4.54 ‰, SD = 1.56) and at Point Bolin (mean = 4.98 ‰, SD = 0.63) were not significantly different ($P \geq 0.05$; table 6-2). The mean δ^{15}N values of *Z. marina* also were relatively low compared to values for macroalgae (fig. 6-2). This relation is consistent with the observation that the low δ^{15}N values for *Z. marina* are associated with groundwater affected by atmospheric inputs (+2 to +8 ‰) or fertilizer inputs (-2 to +2 ‰; McClelland and

Valiela, 1998) rather than from uptake from the water column containing wastewater with δ^{15}N values ranging from +10 to +20 ‰. *Z. marina*, a rooted vascular plant, uses N from the sediment, but may switch to uptake from the water column with increasing N levels in the water column. In contrast, *G. pacifica* and *Ulva* spp. use a holdfast attachment to the bottom and use N from the water column, so they are more likely to reflect wastewater input. For comparison, in Waquoit Bay, Massachusetts, an estuary with 63 percent of the total nitrogen loading contributed by wastewater, *Z. marina* had a mean δ^{15}N value of about 6 ‰ (McClelland and others, 1997; McClelland and Valiela, 1998).

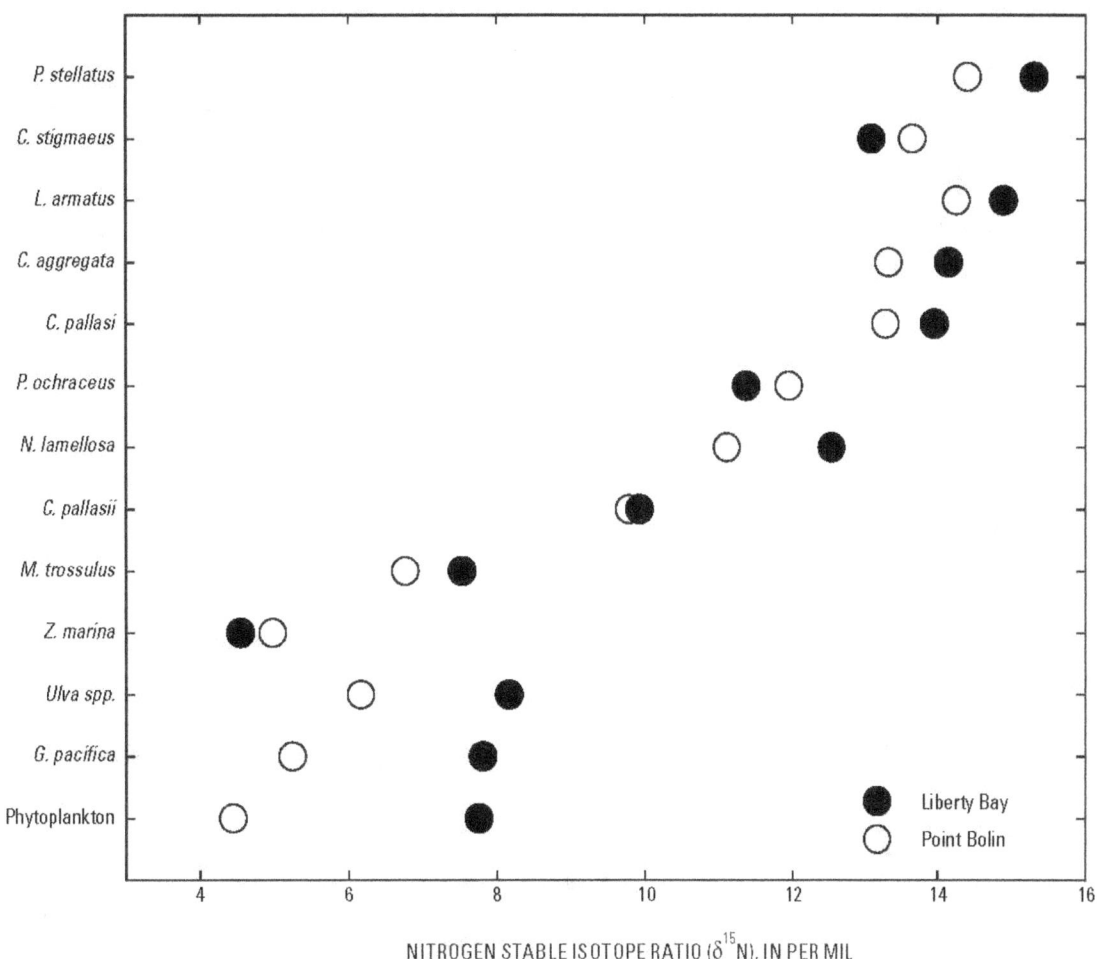

Figure 6-2.   Mean nitrogen isotope (δ15N) values for samples collected in Liberty Bay and at Point Bolin, Central Puget Sound, Washington. See table 6-1 for common names and scientific names.

The mean $\delta^{13}C$ values for *Z. marina* collected in Liberty Bay (mean = -11.39 ‰, SD = 0.28) and at Point Bolin (mean = -9.20 ‰, SD = 0.11) were significantly different ($P < 0.05$; table 6-3). The $\delta^{13}C$ value for *Z. marina* at Point Bolin is close to the -9.1 ‰ (SD = 1.7) cited in a review of $\delta^{13}C$ values for seagrasses (Hemminga and Mateo, 1996). Seasonal variation in $\delta^{13}C$ values for *Z. marina* has been recognized and attributed to differential storage of biochemical components of different isotopic composition (Stephenson and others, 1984; Simenstad and Wissmar, 1985). The $\delta^{13}C$ values at Point Bolin were relatively enriched (difference = +2.19 ‰) compared to Liberty Bay (fig. 6-3). Using light intensity experiments, Grice and others (1996) demonstrated that *Z. marina* subjected to higher light intensity develop less negative $\delta^{13}C$ values. This relation, consistent with higher turbidity of waters in Liberty Bay and less turbid waters at Point Bolin, may have contributed to relatively depleted $\delta^{13}C$ values for *Z. marina* in Liberty Bay. Another alternative is that our samples may be just two points on a gradient of $\delta^{13}C$ values for *Z. marina* that indicates depleted $\delta^{13}C$ values associated with terrestrial C of freshwater creeks and enriched $\delta^{13}C$ values at higher salinities reported by Wozniak and others (2006).

Table 6-3.   Results of a randomization test of mean $\delta^{13}C$ values and 95 percent confidence interval estimated using a bootstrap approach from samples collected in Liberty Bay and at Point Bolin, Central Puget Sound, Washington, 2006.

[**Abbreviations:** CI, confidence interval; ‰, parts per thousand; NA, not applicable]

Sample type	Carbon analysis results				Bootstrapped confidence intervals (1,000 iterations)	
	2.5 percent critical value (‰)	97.5 percent critical value (‰)	Observed difference (‰)	*P* value	Liberty Bay 95 percent CI about the mean (‰)	Point Bolin 95 percent CI about the mean (‰)
Sediment	-0.2937	3.0680	-5.4400	0.0002	(-22.665, -21.241)	(-16.668, -16.310)
Plankton	-2.0820	2.0820	2.0820	0.0450	(-17.938, -16.628)	(-20.760, -18.052)
G. pacifica	-2.4100	2.4100	-3.8700	0.0021	(-17.552, -17.184)	(-13.618, -13.402)
Z. marina	-1.0570	1.1330	NA	<0.0001	(-11.594, -11.138)	(-9.258, -9.130)
Ulva spp.	-0.6840	0.6270	-1.1040	0.0007	(-13.004, -12.656)	(-11.958, -11.464)
P. ochraceus	-0.9840	1.0040	-1.2800	0.0003	(-15.962, -15.212)	(-14.624, -13.940)
N. lamellosa	-0.3300	0.3300	0.2820	0.0939	(-14.380, -14.080)	(-14.680, -14.284)
M. trossulus	-0.5800	0.5800	-0.7320	0.0079	(-17.996, -17.613)	(-17.344, -16.876)
M. secta	-1.6240	1.6240	-1.9120	0.0082	(-16.182, -15.638)	(-15.136, -12.780)
C. stigmaeus	-0.5640	-0.5520	-0.4320	0.0893	(-14.638, -13.858)	(-14.010, -13.670)
L. armatus	-0.7480	0.7520	0.6160	0.1484	(-13.758, -12.646)	(-14.264, -13.524)
C. aggregata	-0.6460	0.6460	0.6860	0.0099	(-14.446, -13.500)	(-14.931, -14.538)
C. pallasii	-1.0080	1.0080	NA	<0.0001	(-16.050, -15.774)	(-17.627, -17.030)
P. stellatus	-0.6060	0.6060	-0.2060	0.5151	(-14.066, -13.214)	(-13.784, -13.132)

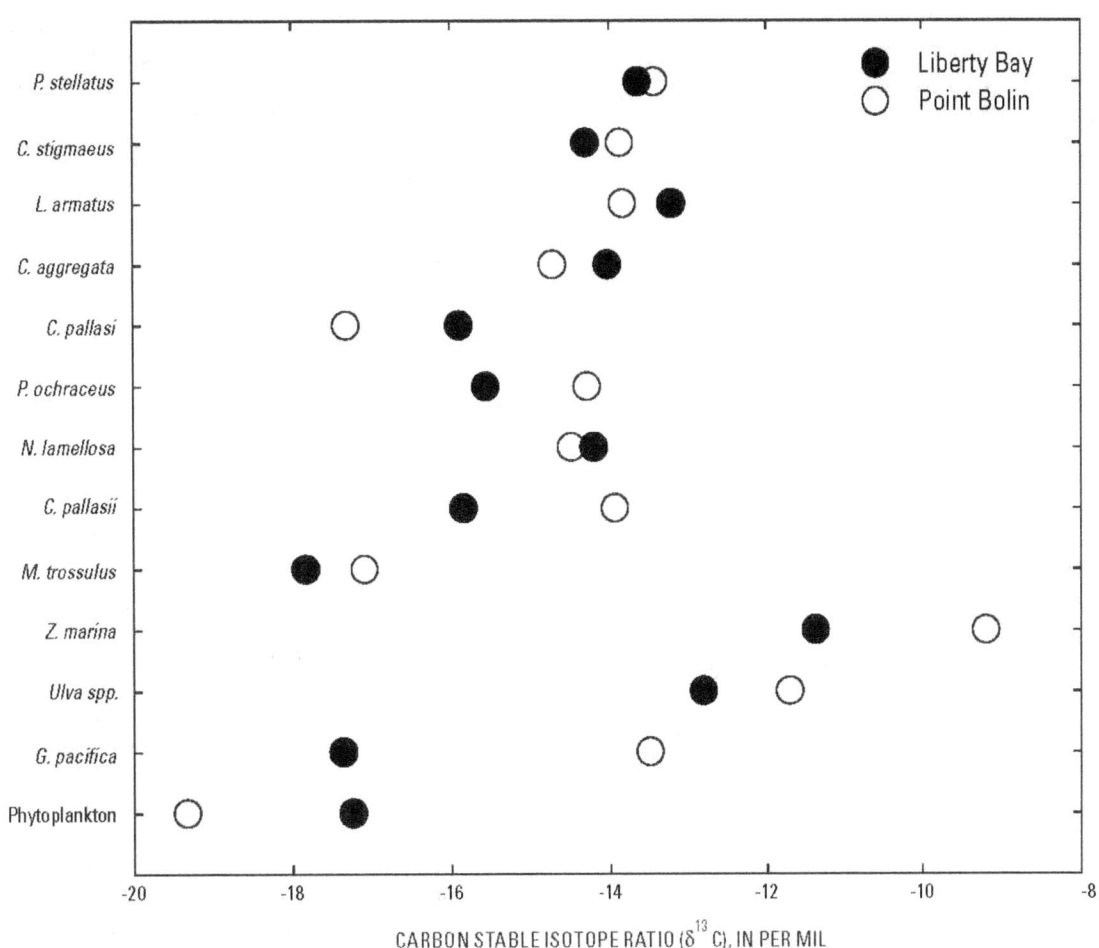

Figure 6-3. Mean carbon isotope (δ¹³C) values for samples collected in Liberty Bay and at Point Bolin, Central Puget Sound, Washington. See table 6-1 for common name and scientific name.

Macroalgae

The mean $\delta^{15}N$ levels in the macroalgae *Ulva* spp. and *G. pacifica* collected in Liberty Bay were significantly higher (difference in means = +2.01 to 2.57 ‰; table 6-1) than in samples collected at Point Bolin ($P < 0.01$; table 6-2). Mean $\delta^{15}N$ values of *Ulva* spp. were 8.17 and 6.16 ‰ for samples from Liberty Bay and Point Bolin, respectively. The mean $\delta^{15}N$ values of *G. pacifica* were 7.81 and 5.24 ‰ for samples collected from Liberty Bay and Point Bolin, respectively. The $\delta^{15}N$ values for *Ulva* spp. collected from three western coast estuaries ranged from 8.9 ‰ in Padilla Bay, Washington, with moderate dissolved inorganic nitrogen (DIN) to 12.5 ‰ in California estuaries, with high DIN (Ruckelhaus and others, 1993; Cloern and others, 2002). In Narrangansett Bay, Rhode Island, the mean $\delta^{15}N$ value of members of the macroalgae family *Ulvaceae* was 9.8 ‰ and Oczkowski and others (2008) concluded that a heavy anthropogenic effect was evident because $\delta^{15}N$ value of Ulvaceae ranged from 7.8 to 7.9 ‰ outside the bay. In marshes on Cape Cod, Massachusetts, Nauset Bay with little urbanization and low groundwater concentration of N, the mean $\delta^{15}N$ value of *Ulva* spp. was 7.1 ‰ (SD = 0.5) (Bannon and Roman, 2008). In Jamaica Bay, New York, where the largest source of freshwater was effluent from six wastewater treatment plants, the mean $\delta^{15}N$ values of *Ulva* spp. averaged 11.6 ‰ (SD = 1.5) (Bannon and Roman, 2008). In a tidal reference site in Massachusetts, *Ulva lactuca* had a $\delta^{15}N$ value of 7.4 ‰ (Wozniak and others, 2006). The mean $\delta^{15}N$ value of *Ulva* spp. of 8.17 ‰ in Liberty Bay generally was greater than values associated with lightly urbanized areas in other studies, but the 6.16 ‰ value for *Ulva* spp. at Point Bolin was less than values reported by others. These results indicate that the level of N in Liberty Bay was relatively enriched, probably with a wastewater origin, in the water column compared to Point Bolin. Because macroalgae use water-column DIN, others have proposed that they are better indicators of N loading than rooted plants (Cole and others, 2004).

The ^{13}C of *Ulva* spp. and *G. pacifica* collected at Point Bolin were relatively enriched in ^{13}C compared to samples collected in Liberty Bay. Differences between the $\delta^{13}C$ values of *Ulva* spp. (mean = -11.72 ‰, SD = 0.32; table 6-1) at Point Bolin and in Liberty Bay (mean = -12.82 ‰, SD = 0.29; table 6-1) were relatively small (1.10 ‰) but statistically significant ($P < 0.001$; table 6-3). The difference between the mean $\delta^{13}C$ values of *G. pacifica* (mean = -13.50 ‰, SD = 0.14; table 6-1) at Point Bolin and in Liberty Bay (mean = -17.37 ‰, SD = 0.24; table 6-1) was larger (3.87 ‰) and statistically significantly ($P < 0.01$; table 6-3). The more

enriched values of $\delta^{13}C$ for *Ulva* spp. compared to *G. pacifica* were consistent with measurements for *Ulva lactuca* (-9.0 ‰) and *Gracilaria tikviahae* (-16.9 ‰) reported by Wozniak and others (2006). The relatively depleted values of $\delta^{13}C$ are consistent with contributions of C from terrestrial sources, including wastewater in Liberty Bay.

Plankton

Based on our observations during collections, our plankton sample probably occurred during a spring bloom of phytoplankton. The $\delta^{15}N$ values of Liberty Bay plankton samples (mean = 7.76 ‰, SD = 2.76) were relatively enriched in ^{15}N compared to the $\delta^{15}N$ values of Point Bolin (mean = 4.44 ‰, SD = 3.14) ($P < 0.05$; table 6-2). The relatively enriched $\delta^{15}N$ values in Liberty Bay were consistent with the bay being identified as at risk for localized eutrophication in Puget Sound (Mackas and Harrison, 1997). However, Liberty Bay $\delta^{15}N$ values were near the mean $\delta^{15}N$ of 8.0 ‰ reported in the San Francisco Bay, California, estuarine system (Cloern and others, 2002). Point Bolin $\delta^{15}N$ values possibly were reduced by the onset of the plankton bloom. Rolff (2000) proposed that during the early part of a bloom in the coastal Baltic Sea, the $\delta^{15}N$ value of cyanobacteria (*Aphanizomenon* spp. and *Nodularia* spp.) was -1.7 ‰, caused by the uptake of gaseous nitrogen through nitrogen fixation. The $\delta^{15}N$ values of plankton in our study were quite variable compared to other biological samples and that may have been caused by tissue of zooplankton in the samples. Inasmuch as phytoplankton blooms are spatially and temporally dynamic, plankton sampling provides limited promise as a long-term monitoring methodology.

Plankton collected in Liberty Bay had a mean $\delta^{13}C$ values of -17.25 ‰ (SD = 1.40) and the mean $\delta^{13}C$ value at Point Bolin was -19.33 ‰ (SD = 1.77). Our values were comparable to collections in the San Francisco Bay estuarine system where the mean $\delta^{13}C$ value of phytoplankton was -21.5 ‰ (range = -26.7 to -17.4 ‰; Cloern and others, 2002). Simenstad and Wissmar (1985) suggested that foam dissolved organic carbon (DOC) excreted by macrophytes and entrained particulate organic carbon (POC) were important sources of C measured in consumers. Further complicating the relations, C available to phytoplankton from DOC and POC, including some phytoplankton, likely is a mixture of C of varied ages and origins (Cloern and others, 2002). Because $\delta^{13}C$ values of plankton were relatively depleted in ^{13}C compared to macroalgae and *Z. marina*, C likely did not originate from those sources.

Bay Mussels

The bay mussel *M. trossulus* was the only suspension filter feeder among the sampled consumers. The mean $\delta^{15}N$ levels of *M. trossulus* were 7.54 ‰ in Liberty Bay and 6.77 ‰ at Point Bolin and were not significantly different ($P > 0.05$; table 6-2). Because it is a filter feeder, *M. trossulus* was compared to plankton collected from the same area; the $\delta^{15}N$ levels of *M. trossulus* were +0.22 to +2.33 ‰ greater in Liberty Bay and at Point Bolin, respectively. A 2–4 ‰ increase was expected for fractionation associated with one trophic level; however, the relation in Liberty Bay fell short of the expected value. An alternate explanation is that *M. trossulus* fed on a mixture of phytoplankton and benthic microalgae associated with the mudflats of Liberty Bay. Others have reported that the diet of the adult common cockle *Cerastoderma edule*, a European mussel, is comprised of as much as 60 percent benthic microalgae resuspended from sediments on mudflats (Sauriau and Kang, 2000; Page and Lastra, 2003). However, the mean $\delta^{15}N$ values for *M. trossulus* collected in Liberty Bay were lower than the $\delta^{15}N$ values for sediments rather than +2 to +4 ‰ higher as expected for a trophic level increase. This relation indicated benthic microalgae resuspended from the mudflats were an unlikely food source for mussels. However, mussels may integrate $\delta^{15}N$ values over a long time period; for example, the ribbed mussel *Geukensia demissa* was estimated to take about 206 days to reach equilibrium in ^{15}N uptake experiments (McKinney and others, 2001). Therefore, it is likely that during the winter months prior to our collection dates the *Mytilus* spp. consumed isotopically light food particulates. Others have reported that phytoplankton or POM have an extended winter minimum in $\delta^{15}N$ values and that may have contributed to relatively low $\delta^{15}N$ levels of mussels in Liberty Bay (Rolff, 2000; Vizzini and Mazzola, 2003).

Filter feeding mussels or clams are good organisms to monitor anthropogenic changes because they are near the bottom of the food web and they integrate phytoplankton and (or) benthic microalgae over time (McKinney and others, 1999; Oczkowski and others, 2008). McKinney and others (2001) reported a positive correlation between $\delta^{15}N$ levels in *G. demissa* in salt marshes of Narragansett Bay, Rhode Island, and the fraction of the watershed in residential land-use. Because the difference in mussel $\delta^{15}N$ levels was relatively small between Liberty Bay and Point Bolin (0.77 ‰ $\delta^{15}N$; table 6-2), it does not support a hypothesis of increased N loading associated with urbanization or septic field leaching.

Because $\delta^{13}C$ is conserved through the trophic levels of a food web, we would expect a consumer such as *M. trossulus* to have $\delta^{13}C$ levels increase ≤ 1 ‰ due to fractionation relative to its food source. The $\delta^{13}C$ levels of *M. trossulus* were -17.84 ‰ (SD = 0.27) in Liberty Bay and -17.10 ‰ (SD = 0.31) at Point Bolin ($P < 0.05$; table 6-3). Although statistically significant, the relatively small differences in $\delta^{13}C$ levels of

M. trossulus suggest they consumed food with similar levels of $\delta^{13}C$ at both of our study sites. We compared *M. trossulus* to plankton collected from the same area and found *M. trossulus* $\delta^{13}C$ levels to be -0.59 ‰ lower in Liberty Bay and +2.23 ‰ higher at Point Bolin. These differences between *M. trossulus* and potential food sources were small compared to differences in $\delta^{13}C$ values ranging from -0.59 to +7.90 ‰ between *M. trossulus* and potential food sources derived from *Z. marina*, *Ulva* spp., and *G. pacifica*. Therefore, the contribution of *Z. marina*, *Ulva* spp., and *G. pacifica* generally could be ruled out as sources of C. Similar to our sampling strategy and findings, Tucker and others (1999) collected macroalgae and blue mussels *M. edulis* nearby, but they also reported that macroalgae contributed little to the particulates ingested by the *M. edulis* in Massachusetts Bay. McKinney and others (2001) studied 10 coastal salt marshes in Narragansett Bay, Rhode Island, reported that the mean $\delta^{13}C$ values for *G. demissa* was -17.6 ‰, and concluded that it derived C primarily from marsh plants and plant litter that differed from the obvious marine sources. Mussel $\delta^{13}C$ values that indicate localized feeding were reported by Ruckelshaus and others (1993) who demonstrated that even in a well-mixed estuary such as Padilla Bay, Washington, $\delta^{13}C$ in mussels corresponded to localized food sources.

Sand Clams

The sand clam *M. secta* is not considered a suspension filter feeder, but rather feeds on detritus and organic matter in or on the substrate (Specht and Lee, 1989). The $\delta^{15}N$ values of *M. secta* were not significantly different between samples in Liberty Bay (mean = 9.93 ‰, SD = 0.75) and at Point Bolin (mean = 9.78 ‰, SD = 0.51) ($P > 0.05$; table 6-2). The $\delta^{15}N$ values of *M. secta* were near the upper end of the range of sediment POM (range 7.46 to 10.33 ‰). Similar to mussels in Liberty Bay, the relatively high levels of $\delta^{15}N$ values in sediments and the relatively low $\delta^{15}N$ values in *M. secta* cannot be accounted for. However, the explanation may be related to observations that *Macoma* spp. can alter their surrounding geochemical environment and may feed selectively on types of POM. Sediment POM was collected near the surface (less than 10 cm) and the *M. secta* from depths of 20–30 cm. At that depth, burrowing of benthic organisms, including *Macoma* spp., may cause changes in the environment that cause gradients in interstitial N compounds due to the nitrification and denitrification at various depths (Grundmanis and Murray, 1977). In addition to changing its surrounding environment, experiments with *Macoma balthica* in Puget Sound indicated it fed selectively on surface sediment particles selecting for protein-coated particles (Taghon, 1982). This selective feeding might result in $\delta^{15}N$ enrichment if the surface phytobenthos and bacteria contributed significantly to the diet.

Similar to several other organisms, $\delta^{13}C$ values of *M. secta* collected in Liberty Bay were significantly depleted compared to samples collected at Point Bolin ($P<0.01$; fig. 6-3). The $\delta^{13}C$ of *M. secta* collected in Liberty Bay (mean = -15.85 ‰, SD = 0.36) and at Point Bolin (mean = -13.94 ‰, SD = 1.52) differed by 1.91 ‰ (table 6-1). This small difference is attributed to the terrigenous sediment POM that likely is more abundant in Liberty Bay sediments. The terrigenous low-density particles high in total organic carbon detected in the sediments of Puget Sound are a potential food source for benthic organisms because of the protein enrichment on the surface of the particles (Taghon, 1982).

Fish

Five species of fish were collected from Liberty Bay and Point Bolin for comparison of their $\delta^{15}N$ values. Although we hypothesized $\delta^{15}N$ values of samples collected in Liberty Bay would be higher than at Point Bolin, we also expected the differences to be minimal because fish integrate $\delta^{15}N$ over a long time and because fish are relatively mobile. However, three of the five fish species collected in Liberty Bay had significantly higher $\delta^{15}N$ values than fish collected at Point Bolin, with differences ranging from -0.56 to +0.66 ‰

($P< 0.05$; table 6-2). The fish with significantly higher $\delta^{15}N$ values in Liberty Bay were the shiner perch *Cymatogaster aggregata*, Pacific herring *Clupea pallasii*, and the starry flounder *Platichthys stellatus*. The *P. stellatus* and staghorn sculpin *Leptocottus armatus* were the highest members of the food web sampled based on relatively high $\delta^{15}N$ values in Liberty Bay and at Point Bolin (figs. 6-4 and 6-5).

In two of the five species of fish, *C. pallasii* and *C. aggregata*, collected in Liberty Bay and at Point Bolin, the mean $\delta^{13}C$ was significantly higher for species in Liberty Bay (table 6-3). The mean $\delta^{13}C$ value of *C. pallasii* collected in Liberty Bay was +1.43 ‰ greater than mean values measured for *C. pallasii* collected at Point Bolin (table 6-1). This was unexpected because of the pelagic nature of *C. pallasii*, but perhaps as juveniles or as an overwintering population they have some site fidelity to Liberty Bay. This difference could be attributed to $\delta^{13}C$ values of phytoplankton samples from Liberty Bay that were +2.08 ‰ greater than mean $\delta^{13}C$ values of phytoplankton collected at Point Bolin. Our results for $\delta^{13}C$ of sediment POM and plankton relative to macroalgae and *Z. marina* support our assumption, based on the biology of these species, that sediment POM and plankton were the primary sources of C for these fish species (figs. 6-4 and 6-5). For the remaining fish species, the mean $\delta^{13}C$ values for samples collected in Liberty Bay and at Point Bolin were not statistically different.

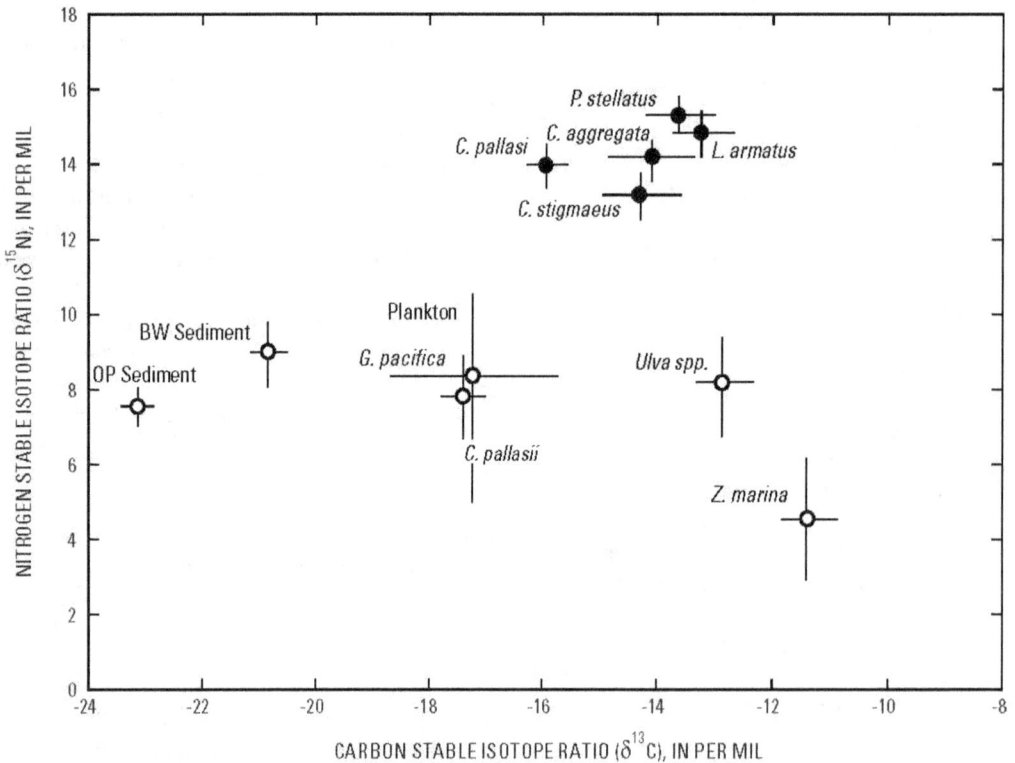

Figure 6-4.   Comparison of mean (±SD) $\delta^{15}N$ and $\delta^{13}C$ values for algae, plankton, sediment particulate organic matter (POM) and fish samples collected in Liberty Bay, Central Puget Sound, Washington. Sediment samples collected near Oyster Plant Park (OP) and the Boardwalk (BW) were collected in Liberty Bay. See table 6-1 for common names and scientific names.

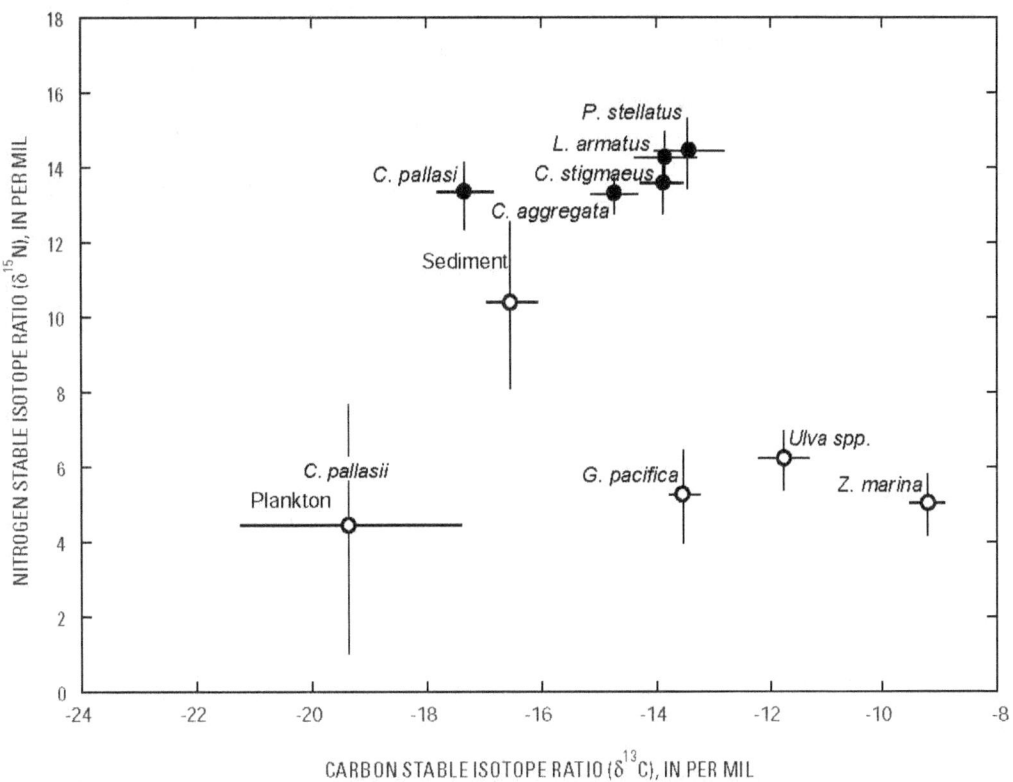

Figure 6-5.   Comparison of mean (±SD) $\delta^{15}N$ and $\delta^{13}C$ values for algae, plankton, sediment particulate organic matter (POM) and fish samples collected at Point Bolin, Central Puget Sound, Washington. See table 6-1 for common names and scientific names.

Summary

Puget Sound is subject to long-term urbanization trends, including increasing human populations, land-use changes, and anthropogenic eutrophication of aquatic ecosystems. These changes usually are accompanied by modifications to the nitrogen and carbon contributions at various trophic levels in the aquatic ecosystem. The analysis of $\delta^{15}N$ and $\delta^{13}C$ values in a study in the Liberty Bay area of central Puget Sound was used to determine if stable isotopes of nitrogen and carbon are potential tools for long-term monitoring. The relatively developed shoreline of Liberty Bay was compared to a less developed reference site on Point Bolin, looking for signs of eutrophication or shifts in food web dynamics.

We report evidence of enrichment of $\delta^{15}N$ values at the primary producer level. The mean $\delta^{15}N$ levels in the macroalgae *Ulva* spp. and *G. pacifica* collected in Liberty Bay were significantly higher than in samples collected at Point Bolin. As suggested in the literature, macroalgae may be suitable primary producers to monitor $\delta^{15}N$ because of the use of nitrogen from the water column. We did not find $\delta^{15}N$ levels in *M. trossulus* or *M. secta* collected in Liberty Bay and at Point Bolin that indicated localized enrichment from wastewater in Liberty Bay; however, the measured $\delta^{15}N$ levels were relatively high, and perhaps indicative of a more widespread enrichment process. In three of the five fish species, the $\delta^{15}N$ values were significantly higher for samples collected in Liberty Bay than for samples from Point Bolin. In general, *P. stellatus* and *L. armatus* had the highest $\delta^{15}N$ values of all samples collected, indicating a relatively high position in the trophic levels that were sampled. However, considerable overlap in $\delta^{15}N$ values among the fish species suggested that nitrogen uptake and use by these fish were similar.

The relatively depleted $\delta^{13}C$ levels reported among lower trophic level organisms in Liberty Bay probably can be attributed to terrigenous inputs from the forested watershed of Liberty Bay. The non-mobile primary producers sampled, including sediment POM, macroalgae, and *Z. marina*, were all relatively depleted in Liberty Bay compared to samples at Point Bolin, suggesting terrestrial inputs were important in contributions to carbon in primary producers. The relative $\delta^{13}C$ levels of fish were depleted compared to samples of macroalgae and *Z. marina*. This relation suggests that $\delta^{13}C$ levels of fish did not result from the expected ≤ 1 ‰ enrichment associated with one or more trophic levels using $\delta^{13}C$ derived from macroalgae and *Z. marina*, but that $\delta^{13}C$ likely was derived from sediment POM and plankton.

Although the use of stable isotopes is appealing in many aspects, the relatively wide ranges in the literature for producers and consumers indicate site specific, spatial, and temporal variation in $\delta^{15}N$ and $\delta^{13}C$ levels. These complex spatial and temporal patterns warrant caution in the use of stable isotopes. Overall, however, we surmise that collecting baseline data and then monitoring some carefully selected primary producers or lower level consumers (such as mussels) through time may be an effective long-term monitoring approach to detect nitrogen enrichment.

Acknowledgments

The authors would like to thank Paul Dorn of the Suquamish Tribe for beach seining operations. We thank Luis Barrantes and Kathleen Byrne-Barrantes of the Liberty Bay Foundation for logistic support and local knowledge. Lisa Gee and Ryan Tomka of the U.S. Geological Survey assisted with specimen collection and sample preparation. We thank our colleagues in the U.S. Geological Survey who supported the Coastal Habitats in Puget Sound (CHIPS) urbanization studies team.

References Cited

Bannon, R.O., and Roman, C.T., 2008, Using stable isotopes to monitor anthropogenic nitrogen inputs to estuaries: Ecological Applications, v. 18, no. 1, p. 22-30.

Cassell, D.L., 2002, A randomization-test wrapper for SAS procedures, in 27th Annual SAS Users Group International Conference, Orlando, Fla., 2002, Proceedings: [USA], SAS Users Group Paper 251-27, p. 1-4.

Cloern, J.E., Canuel, E.A., and Harris, D., 2002, Stable carbon and nitrogen isotope composition of aquatic and terrestrial plants of the San Francisco Bay estuarine system: Limnology and Oceanography, v. 47, no. 3, p. 713-729.

Cole, J.J., Peierls, B.L., Caraco, N.E., and Pace, M.L., 1993, Nitrogen loading of rivers as a human-driven process, in humans as components of ecosystems—The ecology of subtle human effects and populated areas: New York, Springer-Verlag, p. 141-157.

Cole, M.L., Valiela, I., Kroeger, K.D., Tomasky, G.L., Cebrian, J., Wigand, C., McKinney, R.A., Grady, S.P., and Carvalho da Silva, M.H., 2004, Assessment of a $\delta^{15}N$ isotopic method to indicate anthropogenic eutrophication in aquatic ecosystems: Journal of Environmental Quality, v. 33, p. 124-132.

Duggins, D.O., Simenstad, C.A., and Estes, J.A., 1989, Magnification of secondary production by kelp detritus in coastal marine ecosystems: Science, v. 245, p. 170-173.

Ford, J., 1989, The effects of chemical stress on aquatic species composition and community structure, in Ecotoxicology—Problems and approaches: New York, Springer-Verlag, p. 99-144.

Fredriksen, S., 2003, Food web studies in a Norwegian kelp forest based on stable isotope ($\delta^{13}C$ and $\delta^{15}N$) analysis: Marine Ecology Progress Series, v. 260, p. 71-81.

Freyer, H.D., and Aly, A.I.M., 1974, Nitrogen-15 variations in fertilizer nitrogen: Journal of Environmental Quality, v. 3, p. 405-406.

Grice, A.M., Loneragan, N.R., and Dennison, W.C., 1996, Light intensity and the interactions between physiology, morphology and stable isotope ratios in five species of seagrass: Journal of Experimental Marine Biology and Ecology, v. 195, p. 91-110.

Grundmanis, V., and Murray, J.W., 1977, Nitrification and denitrification in marine sediments from Puget Sound: Limnology and Oceanography, v. 22, no. 5, p. 804-813.

Harris, D., Horwath, W.R., and van Kessel, C., 2001, Acid fumigation of soils to remove carbonates prior to total organic carbon or carbon-13 isotopic analysis: Soil Science Society of America Journal, v. 65, p. 1853-1856.

Heaton, T.H.E., 1987, $^{15}N/^{14}N$ ratios of nitrate and ammonium in rain at Pretoria, South Africa: Atmospheric Environment, v. 21, p. 843-852.

Hemminga, M.A., and Mateo, M.A., 1996, Stable carbon isotopes in seagrasses—Variability in ratios and use in ecological studies: Marine Ecology Progress Series, v. 140, p. 285-298.

Kendall, C., Silva, S.R., and Kelly, V.J., 2001, Carbon and nitrogen isotopic compositions of particulate organic matter in four large river systems across the United States: Hydrological Processes, v. 15, p. 1301-1346.

Lake, J.L., McKinney, R.A., Osterman, F.A., Pruell, R.J., Kiddon, J., Ryba, S.A., and Libby, A.D., 2001, Stable nitrogen isotopes as indicators or anthropogenic activities in small freshwater systems: Canadian Journal of Fisheries and Aquatic Sciences, v. 58, p. 870-878.

Mackas, D.L., and Harrison, P.J., 1997, Nitrogenous nutrient sources and sinks in the Juan de Fuca Strait/Strait of Georgia/Puget Sound estuarine system: assessing the potential for eutrophication: Estuarine, Coastal, and Shelf Science, v. 44, p. 1-21.

McClelland, J.W., and Valiela, I., 1998, Linking nitrogen in estuarine producers to land-derived sources: Limnology and Oceanography, v. 43, no. 4, p. 577-585.

McClelland, J.W., Valiela, I., and Michener, R.H., 1997, Nitrogen-stable isotope signatures in estuarine food webs: a record of increasing urbanization in coastal watersheds: Limnology and Oceanography, v. 42, no. 5, p. 930-937.

McCutchan, J.H., Jr., Lewis, W.M., Jr., Kendall, C., and McGrath, C.C., 2003, Variation in trophic shift for stable isotope ratios of carbon, nitrogen, and sulfur: Oikos, v. 102, p. 378-390.

McKinney, R.A., Lake, J.L., Allen, M., and Ryba, S., 1999, Spatial variability in mussels used to assess base level nitrogen isotope ratio in freshwater ecosystems: Hydrobiologia, v. 412, p. 17-24.

McKinney, R.A., Nelson, W.G., Charpentier, M.A., and Wigand, C., 2001, Ribbed mussel nitrogen isotope signatures reflect nitrogen sources in coastal salt marshes: Ecological Applications, v. 11, no. 1, p. 203-214.

Odum, E.P., 1985, Trends expected in stressed ecosystems: Bioscience, v. 35, p. 419-422.

Oczkowski, A., Nixon, S., Henry, K., DiMilla, P., Pilson, M., Granger, S., Buckley, B., Thornber, C., McKinney, R., and Chaves, J., 2008, Distribution and trophic importance of anthropogenic nitrogen in Narragansett Bay—An assessment using stable isotopes: Estuaries and Coasts, v. 31, p. 53-69.

Page, H.M., and Lastra, M., 2003, Diet of intertidal bivalves in the Ría de Arosa (NW Spain)—Evidence from stable C and N analysis: Marine Biology, v. 143, p. 519-532.

Peterson, B.J., and Fry, B., 1987, Stable isotopes in ecosystem studies—Annual Review of Ecology and Systematics, v. 18, p. 293-320.

Rolff, C., 2000, Seasonal variation in $\delta^{13}C$ and $\delta^{15}N$ of size-fractionated plankton at a coastal station in the northern Baltic proper: Marine Ecology Progress Series, v. 203, p. 47-65.

Ruckelshaus, M.H., Wissmar, R.C., and Simenstad, C.A., 1993, The importance of autotroph distribution to mussel growth in a well-mixed, temperate estuary: Estuaries, v. 16, no. 4, p. 898-912.

Sauriau, P.-G., and Kang, C.K., 2000, Stable isotope evidence of benthic microalgae-based growth and secondary production in the suspension feeder Cerastoderma edule (Mollusca, Bivalvia) in the Marennes-Oléron Bay: Hydrobiologia, v. 440, p. 317-329.

Simenstad, C.A., Duggins, D.O., and Quay, P.D., 1993, High turnover of inorganic carbon in kelp habitats as a cause of $\delta^{13}C$ variability in marine food webs: Marine Biology, v. 116, p. 147-160.

Simenstad, C.A., and Wissmar, R.C., 1985, $\delta^{13}C$ evidence of the origins and fates of organic carbon in estuarine and nearshore food webs: Marine Ecology Progress Series, v. 22, p. 141-152.

Singer, G.A., and Battin, T.J., 2007, Anthropogenic subsidies alter stream consumer-resource stoichiometry, biodiversity, and food chains: Ecological Applications, v. 17, no. 2, p. 376-389.

Specht, D.T., and Lee, H., II, 1989, Direct measurement technique for determining ventilation rate in the deposit feeding clam Macoma nasuta (Bivalva, Tellinaceae): Marine Biology, v. 101, p. 211-218.

Stephenson, R.L., Tan, F.C., and Mann, K.H., 1984, Stable carbon isotope variability in marine macrophytes and its implications for food web studies: Marine Biology, v. 81, p. 223-230.

Taghon, G.L., 1982, Optimal foraging by deposit-feeding invertebrates—Roles of particle size and organic coating: Oecologia, v. 52, p. 295-304.

Tucker, J., Sheats, N., Giblin, A.E., Hopkinson, C.S., and Montoya, J.P., 1999, Using stable isotopes to trace sewage-derived material through Boston Harbor and Massachusetts Bay: Marine Environmental Research, v. 48, p. 353-375.

Valiela, I., 1995, Marine ecological processes: New York, Springer-Verlag, 686 p.

Vander Zanden, M.J., and Rasmussen, J.B., 1999, Primary consumer $\delta^{13}C$ and $\delta^{15}N$ and the trophic position of aquatic consumers: Ecology, v. 80, no. 4, p. 1395-1404.

Vizzini, S., and Mazzola, A., 2003, Seasonal variations in the stable carbon and nitrogen isotope ratios ($^{13}C/^{14}C$ and $^{15}N/^{14}N$) of primary producers and consumers in a western Mediterranean coastal lagoon: Marine Biology, v. 142, p. 1009-1018.

Voss, M., Emeis, K.-C., Hille, S., Neumann, T., and Dippner, J.W., 2005, Nitrogen cycle of the Baltic Sea from an isotopic perspective: Global Biogeochemical Cycles, v. 19, GB3001, doi: 10.1029/2004GB002338, accessed October 18, 2010, at http://dx.doi.org/10.1029/2004GB002338.

Voss, M., Larsen, B., Leivuori, M., and Vallius, H., 2000, Stable isotope signals of eutrophication in Baltic Sea sediments: Journal of Marine Systems, v. 25, p. 287-298.

Voss, M., and Struck, U., 1997, Stable nitrogen and carbon isotopes as indicator of eutrophication of the Oder river (Baltic Sea): Marine Chemistry, v. 59, p. 35-39.

Wozniak, A.S., Roman, C.T., Wainright, S.C., McKinney, R.A., and James-Pirri, M., 2006, Monitoring food web changes in tide-restored salt marshes—A carbon stable isotope approach: Estuaries and Coasts, v. 29, no. 4, p. 568-578.

Suggested Citation

Liedtke, T.L., Smith, C.D., and Rondorf, D.W., 2011, Stable isotopes of nitrogen and carbon as tools to monitor eutrophication and trophic dynamics, chap. 6 of Takesue, R.K., ed., Hydrography of and biogeochemical inputs to Liberty Bay, a small urban embayment in Puget Sound, Washington: U.S. Geological Survey Scientific Investigations Report 2011-5152, p. 69-84.

Chapter 7. Spatial Association of Herring Spawn and Shoreline Development in Liberty Bay and Port Orchard, Central Puget Sound, Washington

By Raymond D. Watts[1] and Vivian Queija[2]

Overview

Evidence indicates that populations of Pacific herring are diminishing in Puget Sound and that natural mortality is increasing (Stick, 2005; Stick and others, 2005). Spatially distinct spawning areas are scattered over Puget Sound, Hood Canal, the Strait of Juan de Fuca, and the San Juan Islands (fig. 7-1). The distribution of spawning patches most likely is determined by combinations of factors that are not yet understood, but the absence of spawning in the urbanized main branch of the Sound—from Olympia to Everett—suggests that urbanization may be a factor that suppresses spawning. The urban influence, if present, could be imposed during one or more life cycle stages—eggs, juvenile fish, or adults—and could be related to chemical, physical, or possibly biological alterations that accompany urban development.

The question addressed is whether statistical evidence, in fact, is available indicating that urbanization affects herring spawning. This is a question like many in epidemiology in which investigation requires normalization for multiple factors. At present appropriate indicators—such as ocean currents, physical and biological substrates (sea bottom characteristics), chemical compounds in the water and their concentrations, modification of shoreline structure (for example, bulkhead and pier construction), and biota (for example, forest removal around residences)—for many possible effects are difficult to identify. Some of these factors because of connection and correlation with urbanization deter statistical analysis. In recognition of these complexities, this study does not consider this multitude of potential effects and instead focuses on a single effect: shoreline development.

Humans preferentially settle along shorelines, a phenomenon that is observable at multiple scales: local (Schnaiberg and others, 2002), regional (Xian and others, 2007), and continental (Watts and others, 2007). In Puget Sound and elsewhere, this human inclination leads to earlier development and sustained higher density of development near water than inland. Rather than looking at the broad landscape—at, say, watershed scale—we decided to investigate what seems to be a leading indicator of impacted herring spawning: development immediately adjacent to the waters of Puget Sound.

Study Area

Our study area is Liberty Bay and Port Orchard in central Puget Sound, Washington (fig. 7-2). The study area was specifically selected to capture an urban gradient, which is best expressed between urbanized Poulsbo at the northern end of Liberty Bay and Point Bolin at the southernmost point on the northern shore of Port Orchard. The western and northern shores of the study area are on the Kitsap Peninsula; the eastern shore is on Bainbridge Island.

[1] U.S. Geological Survey, Rocky Mountain Geographic Science Center, 2150 Centre Avenue, MS 516, Fort Collins, CO 80526 (retired).

[2] U.S. Geological Survey EROS Data Center, 909 1st Avenue, 9th Floor, Suite 422, Seattle, WA 98104.

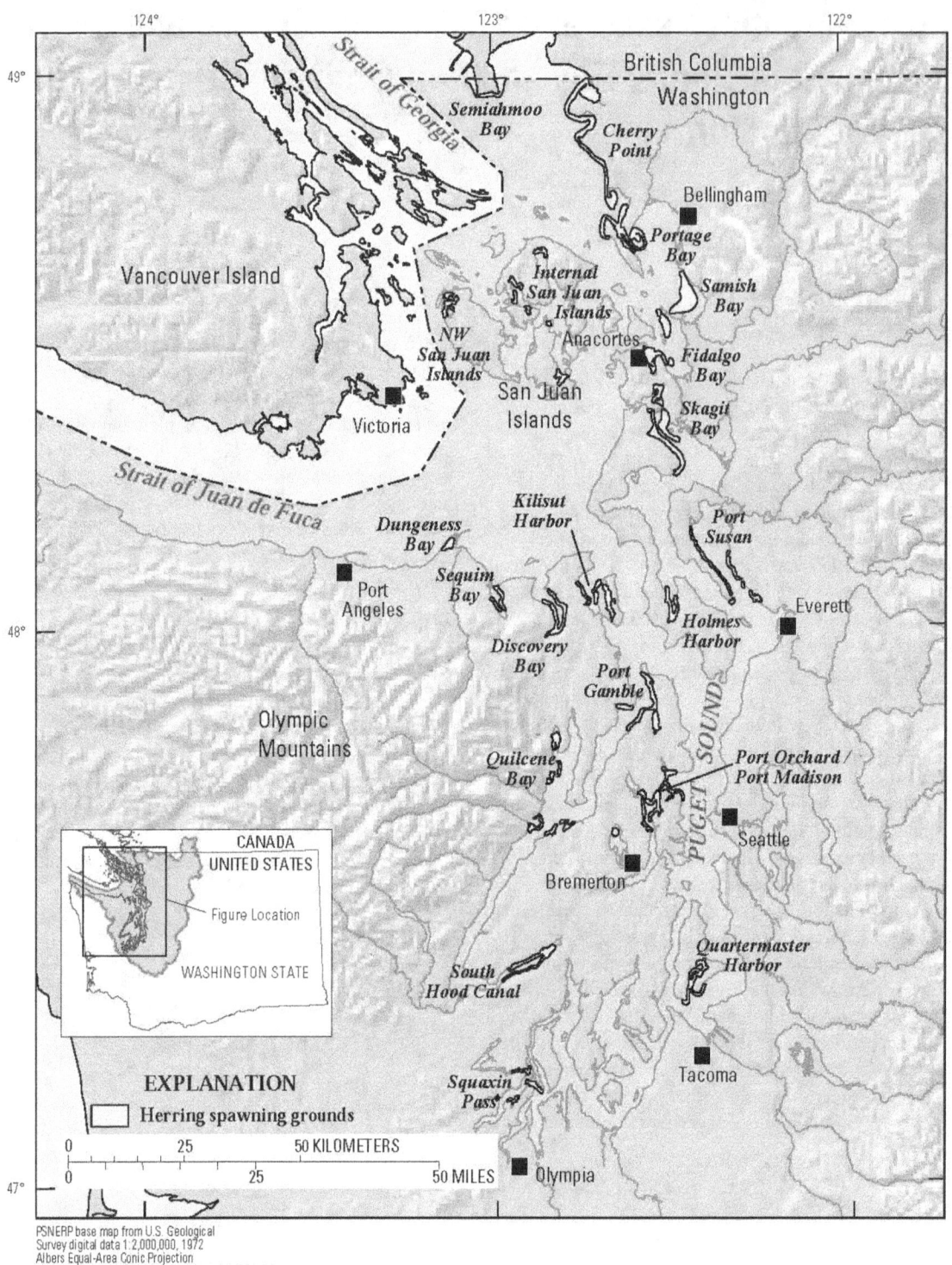

Figure 7-1.   Puget Sound, Washington, and areas of historical herring spawning (after Stick, 2005). The absence of herring spawning grounds along the eastern shore, from Everett to Tacoma, suggests that urban development suppresses herring spawning.

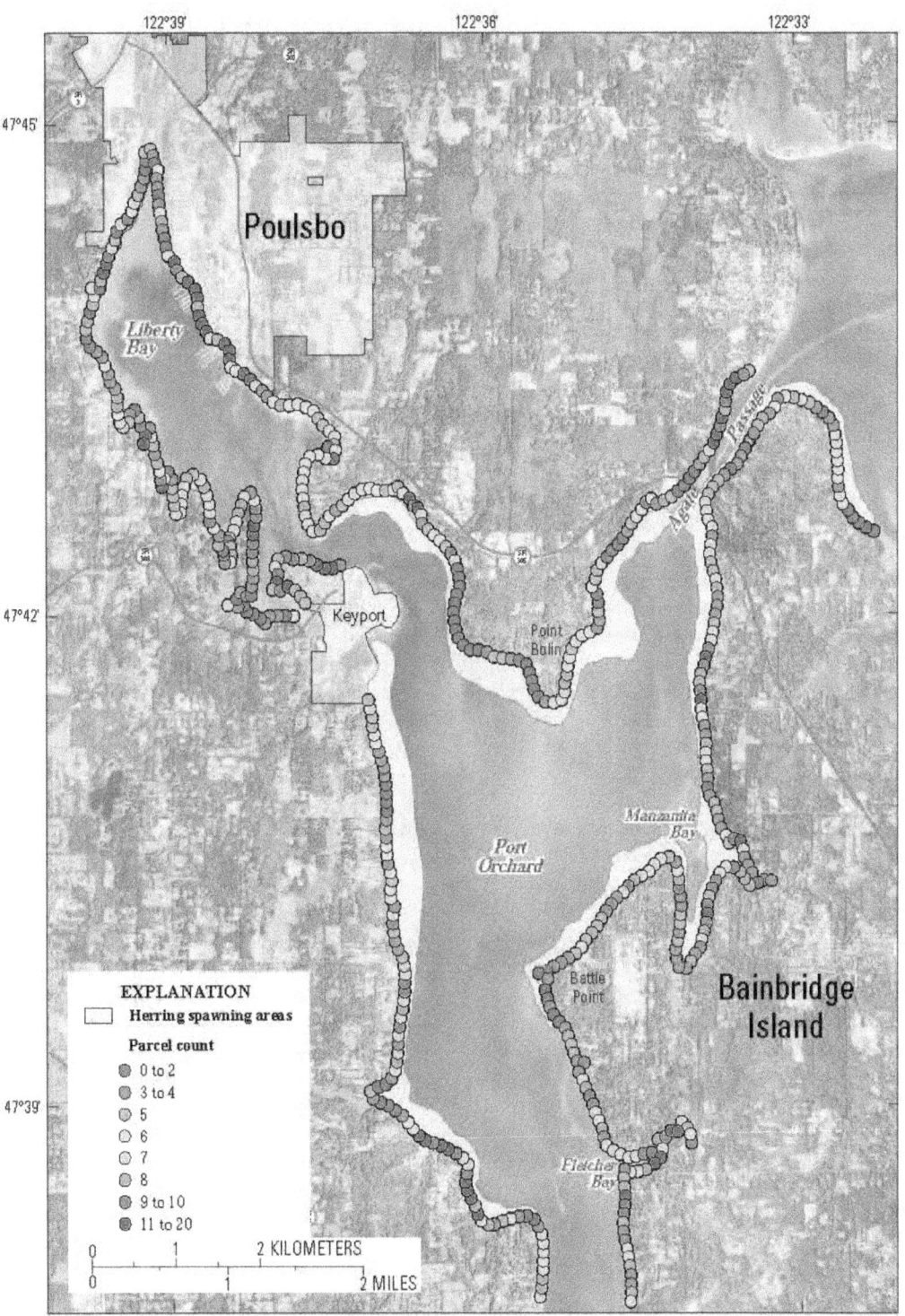

Figure 7-2. Subtidal herring spawning grounds along the shores of Port Orchard and parcel counts along the shores of Liberty Bay and Port Orchard. Parcel counts represent the number of unique owners of land within 100 meters of shore and within a 100-meter circle of the plotted sample point.

Methods

Associations were investigated between an onshore stimulus—urban development—and an offshore response—herring spawn. The two associations were related by establishing sampling points every 100 m along the shoreline, winnowing these as necessary to maintain approximately 100 m spacing where the shoreline is sinuous (fig. 7-2). Sample points on the Keyport industrial site were removed because the degree of urbanization at that site is ambiguous. A total of 577 sample points was selected.

At each point, a Boolean indicator of herring spawn was assigned that expressed occurrence or absence of herring spawn within a 125 m radius. The number of ownership parcels were counted that occur within 100 m of the shoreline near each sample point, with care not to count parcels multiple times when broken by roads and other features with differing ownership. Kitsap County taxation attributes were helpful in preventing multiple counts of broken parcels. High-resolution aerial photography was examined to see what fraction of parcels were developed, and although an exhaustive study was not done, it was determined that in the area examined fewer than 10 percent of shoreline parcels were free of structures, driveways, and other signs of development. As a result of this examination, we inferred that parcel counts are a reasonable proxy for development density. Each sample point was associated with shoreline parcel counts within various radii circles—50, 100, 200, 500, and 1,000 m (1 km).

Houses and other structures that are built along the shoreline are often accompanied by beach modifications—bulkheads, jetties, and piers—that manage erosion and provide access to the water. The Washington Shorezone Inventory (Berry and others, 2001) compiled data about these physical modifications. The summary value of each shorezone unit—a section of shoreline with consistent morphology and materials—expresses the percentage of the unit that has been structurally modified. Each sample point was within a shorezone unit and we attached the summary modification number to each point.

Statistical Analysis

The response variable—presence or absence of herring spawn—is binary, so logistic regression models were used (Collett, 1999) to estimate the probability of observed spawning (P_{spawn}) within 125 m of each of the 577 shoreline sample points.

For each sample point, values for the following parameters were assembled:

- HERSP: 0 or 1, absence or presence of herring spawn
- PD50: the number of shoreline parcels within 50 m
- PD100: the number of shoreline parcels within 100 m
- PD200: the number of shoreline parcels within 200 m
- PD500: the number of shoreline parcels within 500 m
- PD1000: the number of shoreline parcels within 1,000 m
- IN_LB: 0 or 1, location outside or inside of Liberty Bay
- MODPCT: 0 through 100 percent shoreline modification for shorezone unit

Stepwise logistic regression begins by testing the individual predictive power of each explanatory variable. The first single-parameter model tested was shoreline physical modification; no significant ability to predict herring spawn using this model was evident. Deviance —total squared difference between modeled probability and observed values of 1 (spawn present) or 0 (spawn absent)—is used as a summary error measure for logistic regression models; deviance was reduced only from 799.4 to 797.4 by this model.

Our second single-parameter model tested the predictive power of location inside or outside of Liberty Bay. This model reduced deviance from 799 to 420, indicating significant predictive power. Because the properties of Liberty Bay that were creating this strong association with absence of herring spawn are unknown, the data was partitioned and the questions were reformulated.

Location in Liberty Bay is a strong predictor of herring spawning condition. This implies the possibility that Liberty Bay has additional influences that superimpose on those documented. If there are no additional influences, then a model developed using the points outside Liberty Bay should perform well inside Liberty Bay. If, however, there are strong additional influences in the bay, then such a model will perform poorly when applied inside the bay.

One- and two-parameter models were developed based on the 387 points outside Liberty Bay for each of the parcel density search radii (fig. 7-3). Of these models, the 500-m radius combined with the physical modification parameter yielded the highest predictive power, reducing deviance from 420 to 327. The formulas of the logistic regression models are

$$P_{spawn} = \exp(x) / (1 + \exp(x)); \qquad (1)$$
$$x = a + b + PD500 + c \times MODPCT,$$

where a = 2.98, b = –0.05, c = 0.02 for a two-parameter model (minimum of blue, lower curve in fig. 7-3), and a = 3.34, b = –0.04, c = 0 for a one-parameter model (minimum of black, upper curve in fig. 7-3). Application of the two parameter model inside of Liberty Bay predicted that 101 of 190 points would have herring spawn, but 0 points are observed to have it.

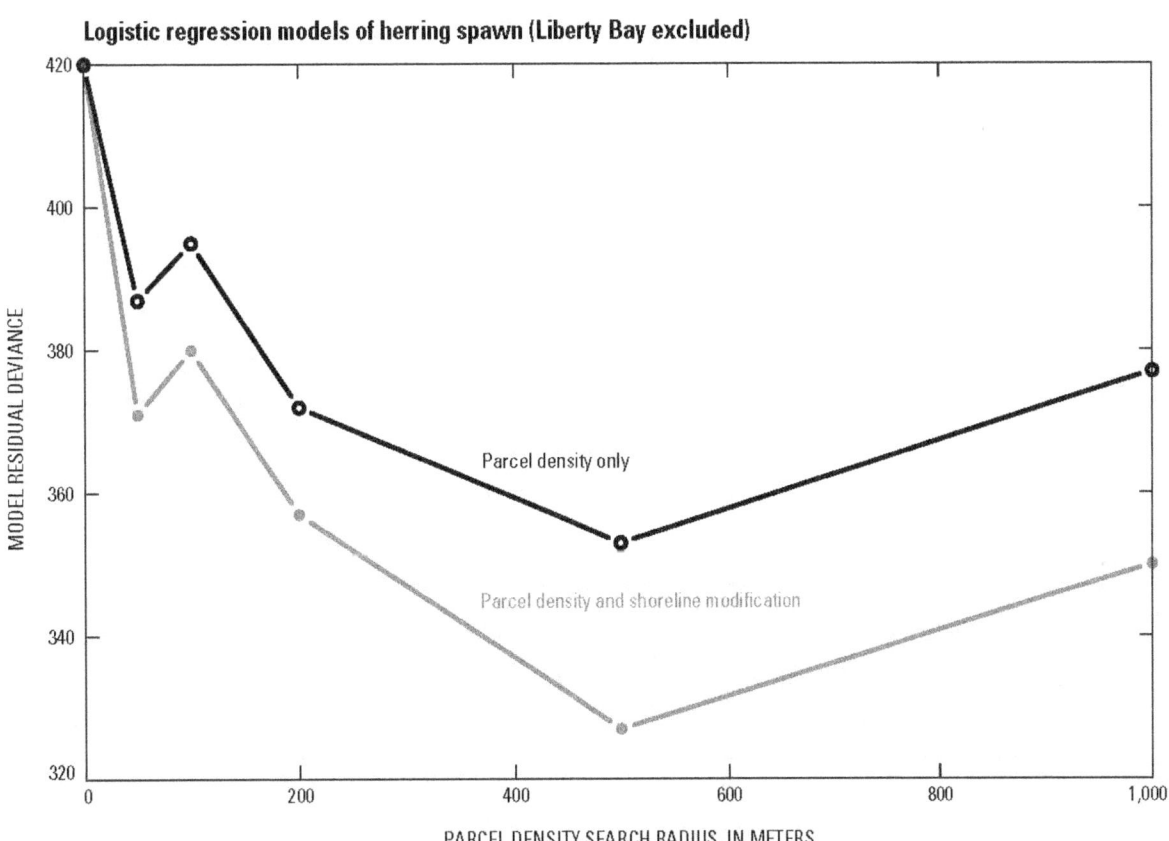

Logistic regression models of herring spawn (Liberty Bay excluded)

Parcel density only

Parcel density and shoreline modification

MODEL RESIDUAL DEVIANCE

PARCEL DENSITY SEARCH RADIUS, IN METERS

Figure 7-3.   Residual deviance in probability models using parcel density (single parameter only) and parcel density and shoreline modification (single parameter and single variable together) as predictive variables. Models were constructed using data only from Port Orchard (excluding Liberty Bay). Parcel densities measured with 500-meter radius provided the best probability predictions.

Discussion

A variety of parcel densities is observed outside Liberty Bay (fig. 7-4). The model applied to the distribution of parcel densities (fig. 7-5, blue line) corresponds well with the no-spawn populations in each class (fig. 7-5, dark bars). The predictive failure for the same model applied inside Liberty Bay supports our conjecture that additional, undocumented factors are present that suppress herring spawn in the Bay. Multiple influences individually or together could account for this suppression of spawn. In chapter 2 of this report, the efficiency of tide driven water exchange between Port Orchard and Liberty Bay is noted. This exchange quickly dilutes chemical contamination, but the vigorous flows also seem to maintain high turbidity in the bay, and could cause herring to avoid the bay for feeding, spawning, or both.

The spawn probability model for Port Orchard (outside Liberty Bay) has a puzzling aspect: the coefficient for shoreline modification is positive, indicating that increasing modification increases the probability of spawning. Eelgrass is the favored spawning substrate, but herring will deposit eggs on jetty rocks, pilings, and other structures when eelgrass is absent or of low quality. In these settings, then, structural modifications may improve spawning habitat. Shoreline conditions that cause erosion, and which are therefore protected by bulkheads and other structures, may provide good herring spawning under some circumstances. An example might be a narrow, steep beach above a sub-tidal bench that supports a healthy eelgrass bed. Thus, we can postulate ways in which structures could either cause or correlate with increased spawning, but we do not presently have sufficient geographic data to test such hypotheses.

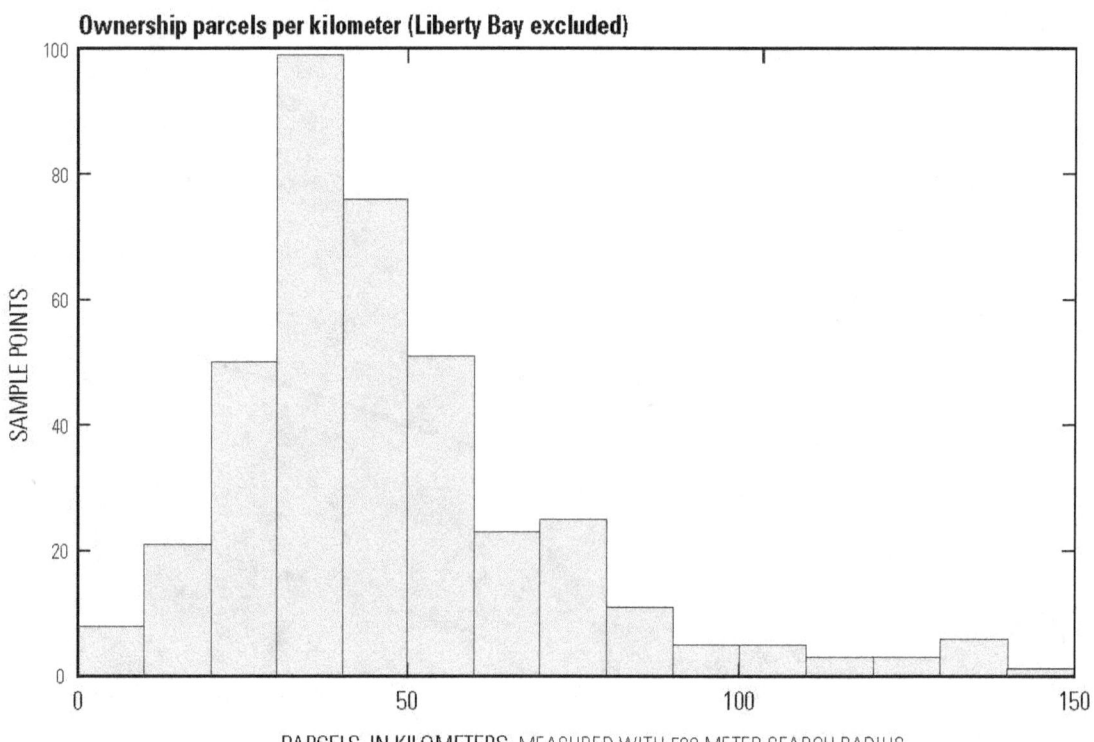

Figure 7-4.   Distribution of observed parcel densities in Port Orchard (Liberty Bay excluded). The functional dependence of probability of herring spawn (or its absence) on parcel density is multiplied by the density distribution to estimate the overall influence of parcel density on herring spawn.

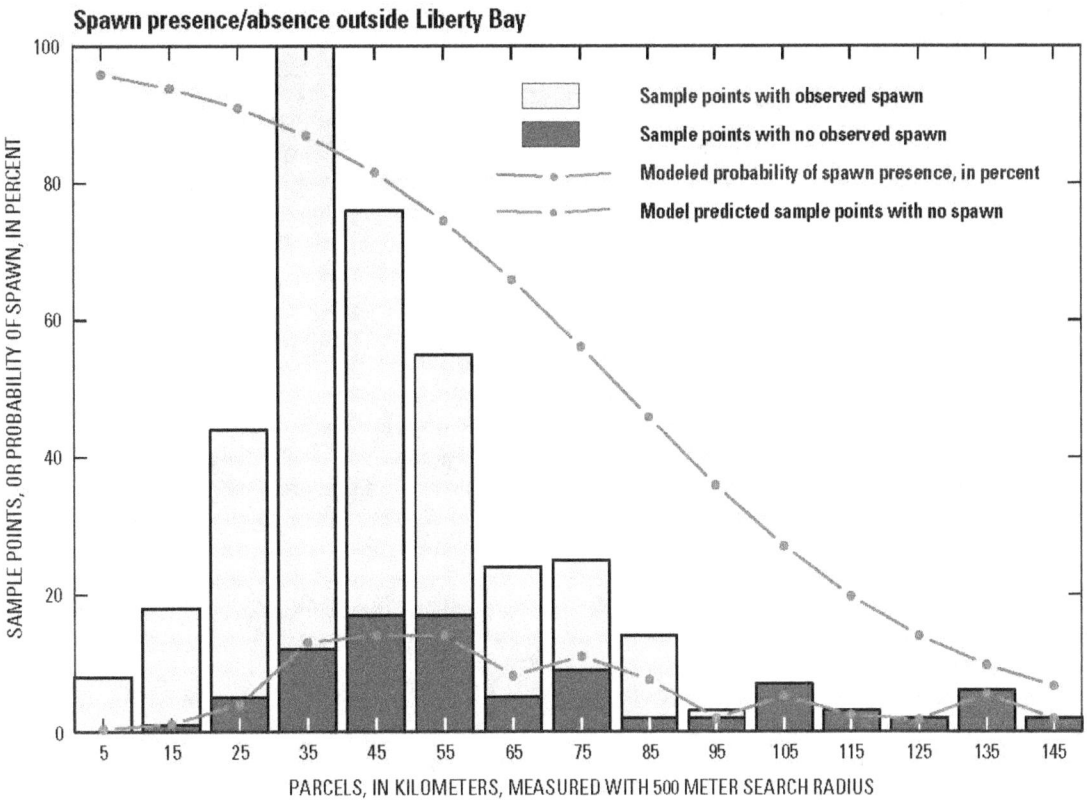

Figure 7-5. Functional probability of herring spawn depending on parcel density.

Parcel densities were measured using circles that spanned as much as 2 km of shoreline (1 km of radius in either direction). This method of measuring parcel density introduces spatial correlations into the densities of nearby points. More sophisticated analysis methods can compensate for the spatial correlation, but we expect changes in model outcome to be modest because Port Orchard herring spawn similarly is spatially correlated; in fact, spawning occurs (or not) over shoreline reaches much greater than the diameters of our sampling circles (fig. 7-2). The best predictor is not the most local, but rather the 500 m radius parcel density measurement. Does this mean that herring actually respond to conditions along 1 km of shoreline? Because herring are mobile creatures, a measurable behavioral response to conditions along a lengthy section of shoreline is quite possible.

Liberty Bay and Port Orchard are a small section of Puget Sound. Within this limited context, our results suggest that the total effect of shoreline development is not favorable to herring spawn. Ownership parcel density, which is a proxy for overall intensity of development, correlates with reduced spawning, whereas physical shoreline modification reduces this suppression. The data also indicate, however, that additional factors are equally strong and that these are at work in Liberty Bay. Multi-parameter modeling of environmental responses to natural- and human-caused environmental conditions is a promising avenue for assessing ecological stresses in Puget Sound. Progress in that direction depends on (1) obtaining data or modeled estimates of a greater variety of condition descriptors and (2) expanding the area of analysis (thus increases the sample size, samples more condition variables, and likely samples different combinations of those variables).

References Cited

Berry, H.D., Harper, J.R., Mumford, T.F., Jr., Bookheim, B.E., Sewell, A.T., and Tamayo, L.J., 2001, The Washington State ShoreZone inventory user's manual: Olympia, Wash., Washington State Department of Natural Resources, 23 p.

Collett, D., 1999, Modelling binary data: Boca Raton, Fla., CRC Press, 369 p.

Schnaiberg, J., Riera, J., Turner, M.G., and Voss, P.R., 2002, Explaining human settlement patterns in a recreational lake district—Vilas County, Wisconsin, USA: Environmental Management, v. 30, no. 1, p. 24-34.

Stick, K.C., 2005, 2004 Washington State herring stock status report: Olympia, Wash., Washington Department of Fish and Wildlife, 86 p.

Stick, Kurt., Costello, Kris, Herring, Chad, Lindquist, Adam, Whitney, Jennifer, and Wildermuth, Darcy, 2005, Distribution and abundance of Pacific Herring (Clupea pallasi) spawn deposition for Cherry Point, Washington stock, 1973–2004, *in* Proceedings of the Puget Sound Georgia Basin Research Conference, March 29–31, 2005: Seattle, Wash., Puget Sound Georgia Basin Research, accessed October 18, 2010, at http://www.engr.washington.edu/epp/psgb/2005psgb/2005proceedings/Papers/A4_STICK.pdf.

Watts, R.D., Compton, R.W., McCammon, J.H., Rich, C.L., Wright, S.M., Owens, T., and Ouren, D.S., 2007, Roadless space of the conterminous United States: Science, v. 316, no. 5825, p. 736–738.

Xian, G., Crane, M., and Su, J., 2007, An analysis of urban development and its environmental impact on the Tampa Bay watershed: Journal of Environmental Management, v. 85, no. 4, p. 965–976.

Suggested Citation

Watts, R.D., and Queija, Vivian, 2011, Spatial association of herring spawn and shoreline development in Liberty Bay and Port Orchard, Central Puget Sound, Washington, chap. 7 *of* Takesue, R.K., ed., Hydrography of and biogeochemical inputs to Liberty Bay, a small urban embayment in Puget Sound, Washington: U.S. Geological Survey Scientific Investigations Report 2011–5152, p. 85-92.

Chapter 8. Synthesis

By Renee K. Takesue[1], Richard S. Dinicola[2], Jessica R. Lacy[1], Theresa L. Liedtke[3], Dennis W. Rondorf[3], Collin D. Smith[3], and Raymond D. Watts[4]

Introduction

Nearshore environments of Puget Sound are increasingly disturbed by human activities. Such non-natural disturbances can alter physical, chemical, and biological conditions and processes, and impair ecological functions. A comparative approach between a semi-urbanized site (Liberty Bay) and a non-urbanized site (Point Bolin) was used to determine whether certain attributes of the nearshore ecosystem were associated with urbanization. An embayment was selected as the semi-urbanized site because it was expected that the greater retention of materials in an embayment compared to the open shore would maximize the likelihood of detecting anthropogenic contaminants and effects. An initial assessment of the physical, chemical, and biological characteristics of Liberty Bay and Point Bolin comprised most of the study. Anthropogenic inputs from the urban watershed were identified and quantified, as well as potential ecological effects. A statistical model related degrees of shoreline development with the probability of forage fish spawn. The focus on a semi-urbanized site indicates the likelihood that future population growth in the Puget Sound basin will give rise to concentrated nearshore residential development rather than large urban or industrial centers.

Nearshore Characteristics

Physical Environment

Several characteristics related to the physical energy environment differed between Liberty Bay and Point Bolin. These included current speeds, the retention of materials, bottom sediment grain-size distributions, and total suspended sediment concentrations. Although Liberty Bay is relatively

[1] U.S. Geological Survey Pacific Coastal and Marine Science Center, 400 Natural Bridges Drive, Santa Cruz, CA 95060.

[2] U.S. Geological Survey Washington Water Science Center, 934 Broadway, Tacoma, WA 98402.

[3] U.S. Geological Survey Western Fisheries Research Center, Columbia River Research Laboratory, 5501-A Cook-Underwood Road, Cook, WA 98605.

[4] U.S. Geological Survey, Rocky Mountain Geographic Science Center, 2150 Centre Avenue, MS 516, Fort Collins, CO 80526 (retired).

protected from waves, its relatively large tidal prism and shallow depths produce significant tidal currents. Tidal currents in the main channel of Liberty Bay reach 30 cm/s and are stronger at the constriction near Lemolo. Tidal currents are substantially weaker between Point Bolin and Keyport and exceed 100 cm/s in Agate Passage. For most of Liberty Bay, water residence time during spring tides was estimated to be less than a tidal cycle. Residence time is, of course, greater in coves and during neap tides. Although dilution and mixing certainly are less in Liberty Bay than at open shore sites in Puget Sound, the exchange of materials between Liberty Bay and Port Orchard nevertheless is quite high.

During April–May 2006, surface salinity rarely differed by more than 1 practical salinity unit (psu) between the Liberty Bay mooring site (LB) and the Point Bolin mooring site (PB) despite freshwater input to Liberty Bay from several creeks. Occasions when salinity at LB was more than 0.5 psu lower than at PB lasted less than a day, providing further evidence that water frequently exchanges between Liberty Bay and Port Orchard. Tidal variations in surface salinity were less than 1 psu at LB, and even less at PB, indicating that spatial gradients in salinity were weak. Because of the lack of significant salinity and temperature gradients it was concluded that gravitational circulation was not an important component of transport relative to tidal currents during this period.

The relatively protected nature of Liberty Bay, combined with the input of fine sediment from the watershed, contributes to the deposition and retention of fine sediment and organic matter. Fine sediment is more susceptible to resuspension than sand, contributing to turbid conditions in the bottom waters of Liberty Bay, as was observed in the deepest part of the bay (LB). Even at the surface, LB was consistently more turbid than PB. Suspended sediment concentration (SSC) at LB varied tidally, with peaks occurring at the beginning of the flooding tide. This phasing of elevated SSC and tidal currents serves to retain sediment within the bay, as resuspended sediment is preferentially transported in the flood direction (into the bay). Fine sediment in Liberty Bay also was chemically reducing, or anoxic. Anoxia and hydrogen sulfide, a byproduct of microbial respiration in reducing sediment, can be harmful to benthic organisms. Thus, muddy, low-energy areas of Liberty Bay are less-favorable habitat for certain benthic organisms than sandy, higher-energy environments.

In Liberty Bay and Port Orchard nearly all ownership parcels have been developed. Development alters physical conditions of the shore, both physical and chemical conditions near the shore, and beach ecology (MacDonald and others, 1994; Thom others, 1994). Two indicators of direct physical modification were used to quantify degrees of development. The density of parcels within 40 m of the shoreline was used as a surrogate for and composite of more direct measurements of physical modifications (for example, impervious surface area, density of structures, and vegetation canopy removal). Estimates of percentage shoreline modification based on constructed bulkheads, piers, jetties, and so on per shore unit (a unit is an unbroken stretch of shoreline with similar materials and morphology) were obtained from the Washington ShoreZone Inventory (Washington State Department of Natural Resources, 2010), at http://fortress. wa.gov/dnr/app1/dataweb/dmmatrix.html. Shorelines in the city of Poulsbo and the rest of Liberty Bay were six times and three times more densely developed, respectively, than the shoreline on the western side of Point Bolin. Degrees of shoreline modification were five times greater in the city of Poulsbo and four times greater around the rest of Liberty Bay compared to the western side of Point Bolin.

Biological Characteristics

Several phytoplankton blooms occurred at sites LB and PB during April and May 2006 along with decreases in dissolved nutrient concentrations. Nutrients decreased in surface waters before they decreased at depth, which is evidence that nutrient uptake by phytoplankton caused low nutrient concentrations. No difference in bloom intensity was detected between the two sites, although detection was limited by uncertainty in the field calibrations. Bloom concentrations at both sites were typical of those observed at enclosed Puget Sound sites, which are greater than concentrations observed at open water sites (Newton, and others, 2002). However, it is not clear that elevated phytoplankton concentrations at enclosed sites are caused by anthropogenic nutrient inputs because decreased vertical mixing also contributes to phytoplankton growth (Newton and others, 2002). The amount of nitrogen required to support standing stocks of phytoplankton in Liberty Bay, based on a yield of about 2 μg/L chlorophyll a per 1 μmol/L dissolved inorganic nitrogen (Gowen and others, 1992; Edwards and others 2005), was between 2 to 35 times more than was measured in the water column in April and May 2006. Such nitrogen requirements could have been satisfied by marine nutrients.

There were no eelgrass beds inside Liberty Bay north of the constriction at Keyport–Lemolo. Fringing beds occur on the western side of Point Bolin and in an extensive meadow on the eastern side of Point Bolin at Sandy Hook. The main difference in water quality characteristics between Liberty Bay and Point Bolin was higher turbidity in Liberty Bay. This may be an indication that turbidity limits eelgrass growth and re-establishment in Liberty Bay. Eelgrass seedlings are more susceptible to environmental stressors than are healthy mature plants, so even if seeds were transported into Liberty Bay, seedling establishment could fail because of sub-optimal light levels. Low oxygen concentrations and hydrogen sulfide also impair eelgrass growth and could contribute to seedling failure. We do not known if eelgrass has ever grown in Liberty Bay, and we did not investigate whether healthy mature plants could grow in Liberty Bay if they were transplanted there.

Contaminant Inputs To Liberty Bay From The Urbanized Watershed

Contaminants enter the nearshore dissolved in surface water and groundwater, bound to terrigenous particles, and by atmospheric deposition. Preceding and during this study, there were three sewage spills into Liberty Bay from an aging municipal sewer line, and most of the tributary creeks draining residential areas with septic systems have histories of elevated fecal coliform levels (Kitsap County, 2006). Thus, compounds associated with wastewater were expected to be detected in Liberty Bay. Two types of contaminants were measured directly in freshwater entering Liberty Bay: nutrients in groundwater, and pharmaceutical and personal care products (PPCPs) in surface water and groundwater. Occurrences and magnitudes of wastewater indicator compounds and wastewater nitrogen indicators at the head of Liberty Bay compared to less-developed sites suggest inputs that are a result of urbanization.

Inputs of Pharmaceutical and Personal Care Products in Creeks and Groundwater

Pharmaceutical compounds distinctive of household and human waste streams were detected in creeks and groundwater flowing into Liberty Bay from watersheds served by on-site septic systems. Septic systems are not designed to treat for PPCPs, so surface runoff and shallow groundwater in residential watersheds served by on-site septic systems may be a widespread source of PPCPs to Liberty Bay and the broader Puget Sound nearshore.

Inputs of Nitrogen and Coastal Groundwater Discharge

Submarine groundwater discharge (SGD) has been shown to affect nearshore material budgets and play a crucial role in the ecological wellbeing of some coastal systems. Such fluxes may be particularly important in areas of Puget Sound where the physical drivers that control SGD (for example, tidal range, recharge, geology) support enhanced land/sea exchange. Although the groundwater that discharges into the sea often consists of natural geologic weathering products, anthropogenic activities in watersheds possibly can introduce nutrients, trace elements, fertilizers, and organic compounds that eventually filter into groundwater.

The dissolved inorganic nitrogen (DIN) concentration measured in shallow groundwater at Oyster Plant Park in Liberty Bay was more than five times greater than the DIN concentration measured at Sandy Hook, a less densely populated site on the eastern shore of Point Bolin, suggesting more anthropogenic nitrogen migrates to groundwater from more densely populated areas. The DIN concentrations measured in groundwater at both Oyster Plant Park and Sandy Hook were two to five times greater than DIN concentrations measured in seawater, which indicates that SGD is a source of nutrients to marine waters of Liberty Bay and Port Orchard. Although the estimated SGD rate at Sandy Hook was about two times greater than in Liberty Bay, higher DIN concentrations in Liberty Bay groundwater resulted in an estimated DIN flux in SGD that was more than three times greater in Liberty Bay than at Sandy Hook. The estimated nitrogen fluxes to Liberty Bay and Port Orchard were slightly less than fluxes estimated to Lynch Cove in Hood Canal, where upland land-use is similar to that around Point Bolin. The nitrogen fluxes at all three of these sites, in turn, were about seven times lower than at Skagit Bay, where land use is predominately agricultural. However, comparable estimates of nitrogen fluxes in Puget Sound SGD are too few to determine whether the fluxes measured in Liberty Bay and at Point Bolin were affected by urbanization.

Contaminants in Nearshore Ecosystem

Wastewater Indicator Compounds in Sediment and Water

Associations of wastewater indicator compounds with certain commercial, residential, or agricultural activities allowed types of land uses to be inferred that contributed chemical loadings and pathways of contaminant transport into Liberty Bay. Metals and polycyclic aromatic hydrocarbons (PAHs) in Liberty Bay bottom sediment show the importance of commercial waste streams and runoff from the land surface, particularly roads, around the city of Poulsbo. Household wastewater compounds such as deodorants and disinfectants were not detected in Liberty Bay sediment, but caffeine and ibuprofen, which are used only by humans, were detectable in creeks and groundwater flowing to Liberty Bay. The discrepancy in the detection of household compounds in sediment and water is related to their differing behaviors in the environment: their tendency to concentrate on particles, their susceptibility to degradation, and their dilution with seawater below levels of detection. This highlights the importance of analyzing water-borne and sediment-bound contaminant fractions to get a more complete picture about human activities and land uses that introduce chemicals into Liberty Bay.

Anthropogenic Nitrogen in Water and Biota

Runoff from urban areas and wastewater effluent can contribute large fractions, as much as 20 percent and 81 percent, respectively, of nitrogen loads to estuaries and coasts (Driscoll and others, 2003). The Washington State Department of Ecology has identified high ammonium (NH_4^+) concentrations to be indicative of wastewater (sewage) nitrogen input (Newton and others, 2002). An NH_4^+ concentration typical of oceanic water (Admiralty Inlet) is about 2 µM (0.03 mg/L) and an NH_4^+ concentration around 5 µM (0.07 mg/L) is considered 'high' (Newton and others, 2002). Based on this criterion, bottom water in Liberty Bay in late April and early May (3 and 4 µM, respectively) could have been slightly affected by wastewater nitrogen.

Nitrogen from wastewater (anthropogenic nitrogen) can be identified in nearshore water, sediment, and biota based on nitrogen stable isotope ratios ($\delta^{15}N$) (McClelland and Valiela, 1998; Lake and others, 2001; Cole and others, 2004). Enrichments of 2 per mil in $\delta^{15}N$ of sedimentary particulate organic matter and macroalgae in Liberty Bay compared to enrichments at the Point Bolin reference site were evidence of nitrogen originating from wastewater. Macroalgae use a holdfast attachment to hard-bottom substrate and obtain N entirely from the water column. Thus, the $\delta^{15}N$ value of macroalgae reflects the $\delta^{15}N$ composition of water, making them good indicators of N loading (Cole and others, 2004). In addition to macroalgae, filter-feeding mussels can be used to monitor anthropogenic-N loading because they are near the bottom of the food web and integrate changes in the $\delta^{15}N$ compositions of phytoplankton and (or) benthic macroalgae over time (Lake and others 1999; McKinney and others, 1999; Oczkowski and others, 2008). The mean $\delta^{15}N$ value of Liberty Bay blue mussels (7.5 per mil) was higher than Point Bolin mussels (6.8 per mil), consistent with anthropogenic nitrogen inputs, but the difference was not statistically significant because of large $\delta^{15}N$ variability among individual mussels. Values of $\delta^{15}N$ hold potential as a long-term monitoring tool to investigate anthropogenic nitrogen loading; however, site-specific, spatial, and temporal variability can be significant and must be accounted for in the study design. Carefully selected primary producers or low-level consumers are considered the best candidates for long-term monitoring efforts.

Ecological Effects

Eutrophication, or the stimulation of excessive plant growth, can arise from anthropogenic nutrient loading (Driscoll and others, 2003). Eutrophication leads to algal blooms, degradation of water quality associated with turbidity and anoxia, and loss of species that shortens food chains and decreases species richness (Woodwell, 1983; Odum, 1985; Ford, 1989). Evaluations of trophic relations using stable isotopes did not show an altered food chain in Liberty Bay compared to Point Bolin. This result could be affected by a number of factors, including the degree of anthropogenic nitrogen-loading into Liberty Bay at the time of the study, the timing of the field program, and the fact that stable isotopes were only measured in species that were present at both sites. Stable isotope values in co-occurring species alone were not indicative of shortened food chains arising from environmental stressors. In retrospect, it appears that a combination of stable isotope values and biological assessments of species richness would have helped determine whether food chains differed at the urbanized and non-urbanized sites.

Ecological health risks from metals and some organic contaminants such as PAHs are well known. Metal toxicity may result in liver or kidney damage in invertebrates, fish, and wildlife (Eisler, 1985), whereas certain PAHs can cause cancers, genetic mutations, and birth defects (Douben, 2003). Because metals and PAHs accumulate in sediment, they are a particular risk to organisms that live in, on, or feed on nearshore sediment such as bottom fish, deposit-feeding invertebrates, and rooted aquatic plants. The concentrations of organic contaminants and PPCPs that remain dissolved in water are several orders of magnitude lower than concentrations in sediment. Such low concentrations are not likely to cause acute toxicity in aquatic organisms, but constant use and discharge of PPCPs to Liberty Bay could lead to chronic, low-level exposure of biota. Sub-lethal levels of antibiotic, anti-inflammatory, antiepileptic, and beta-blocking drugs, all of which were detected in Liberty Bay surface and groundwater (Dougherty and others, 2010), variously cause hormonal imbalances, reduced fertility, retarded growth, and photosynthesis inhibition in aquatic organisms (Khetan and Collins, 2007). The likelihood of population growth, population aging, and increasing household and pharmaceutical product use around Liberty Bay raises concern about inputs and effects of these compounds on aquatic ecosystems.

The greater effort in this study focused on characterizing the natural physical environment, potential effect mechanisms, and ecological responses. Another study component directly tested for statistical evidence of an association between human-induced change—shoreline development and shore physical alteration—and an important ecological response— herring spawn. Association does not demonstrate cause and effect, but lack of statistical association strongly suggests lack of cause and effect connections. The logistic regression analysis showed that shoreline development density is a strong predictor for suppression of herring spawn. The model, however, underestimates the amount of suppression in Liberty Bay, which suggests that additional parameters are needed to explain spawning conditions in the environment of the bay. Curiously, the model also shows a slight decrease of spawn suppression as shoreline modification increases. Mechanisms that could cause shoreline modifications to improve spawning habitat at the time of this study are not known.

Evaluation of the Liberty Bay Study

The goals of this study were (1) to describe the physical and biogeochemical characteristics of Liberty Bay and Point Bolin, (2) to identify inputs of urban chemicals to the nearshore, and (3) to identify altered nearshore habitats and (or) processes arising from urbanization. The greatest challenge was to achieve meaningful progress toward the study goals in the limited context of a proof-of-concept (pilot) study. Because the first goal was to make connections between physical, chemical, and biological processes, a short, field-intensive study was conducted rather than a longer less-intensive study. Measurements were made over 2 months in spring, but understanding of nutrient and plankton dynamics and eutrophication would have greatly benefited from year-round monitoring because many nearshore and ecological processes vary seasonally. For example, stream discharge and stormwater runoff increase in autumn and winter when rainfall increases. Plant growth peaks in late spring and summer in response to increasing solar radiation. Eutrophic conditions can arise in late summer or fall when thermal stratification of the water column inhibits vertical mixing. Thus, conclusions about Liberty Bay water quality in April and May cannot be readily extrapolated to other seasons.

Several types of urban contaminants were identified in freshwater flowing to Liberty Bay and in nearshore sediment and biota. Among these were household pharmaceutical and personal care products, compounds associated with roads and vehicle operations (PAHs), industrial metals, and wastewater nitrogen. This study showed the importance of non-point sources (for example, septic systems, urban runoff, submarine groundwater discharge) in the delivery of contaminants to the nearshore. The absence of eelgrass beds inside Liberty Bay could not be ascribed uniquely to urbanization because physical characteristics of the bay also affect habitat quality. Additionally, ecological effects associated with wastewater nitrogen inputs could not be definitively identified. A statistical model (without effect mechanisms) showed that parcel density was a strong predictor of herring spawn suppression although the presence of shoreline structures was associated with a slight increase in herring spawn.

Effects of urbanization were investigated in an embayment, based on the premise that the signal of semi-urbanization would be stronger in an enclosed water body than in an open water environment. It was assumed that, as a consequence of longer residence time, the spatial scale of the urban effects would be large relative to the spatial scale of transport. This was not the case in Liberty Bay, where tidal flushing transported water and dissolved constituents from the more-urbanized head of the bay to the less-urbanized mouth of the bay daily. In larger, deeper embayments with a greater degree of shoreline urbanization, spatial scales of urban effects and tidal flushing likely are more independent. Differences in anthropogenic effects may be evident at the scale of the study, such as those suggested by the elevated nitrogen concentrations in shallow groundwater in the more urbanized part of the study area, but given the level of mixing, detection requires relatively sensitive methods and (or) longer periods of study. This highlights the difficulties in determining the effects of anthropogenic inputs in Puget Sound, where mixing processes are spatially variable, semi-urban inputs are ubiquitous, and patterns of urbanization are complex, with gradients at different scales superimposed on each other. The selection of spatial scale is not straightforward and is critical to the success of the study. A study approach targeting larger scales and using many sites likely would be more successful.

The identification of a reference site with which to compare characteristics of an urbanized nearshore was problematic because the suitability of a site for a certain suite of parameters, such as contaminants, necessarily was not the same as for a different suite of parameters, such as the physical energy environment. Physical and water-quality parameters (temperature, salinity, turbidity, fluorescence, nutrients, oxygen, pH, chlorophyll) were measured in the deepest part of Liberty Bay and on the western side of Point Bolin. These sites exchanged water over a tidal cycle, and thus were not independent. Metals, $\delta^{15}N$ values, and PPCPs were compared between the urbanized head of Liberty Bay and the eastern side of Point Bolin (Sandy Hook). Little or no exchange of materials occurred between these sites, so the eastern side of Point Bolin was an appropriate reference site for metals and $\delta^{15}N$ values; however, it was not an appropriate reference site for PPCPs because many of these compounds originate from household waste streams and because shoreline land-use around Point Bolin primarily is residential. Finally, no reference site was identified for sedimentary wastewater indicator compounds. Both sediment samples for wastewater indicator compound determinations were collected inside Liberty Bay, but at sites with differing degrees of urbanization. Instead of a reference site, we assumed that sediment at a site without wastewater inputs would contain no wastewater indicator compounds, and therefore that the presence of wastewater indicator compounds in Liberty Bay sediment was indicative of wastewater inputs.

In summary, the 2 month-long study of Liberty Bay was successful as a pilot study despite the limited understanding gained about seasonally varying processes. Gains in scientific and logistic understanding allow prioritization of study components such as duration, intensity, analytes, selection of reference sites, and spatial scales for future studies. A study area approximately 20 km alongshore was an appropriate scale for a team of 8 to 10 scientists for 2 months of 1 year. As is the intent of pilot studies, initial field observations gave rise to study design refinements and new research questions that could be pursued in subsequent years.

The broad perception is that ecological conditions in Puget Sound are deteriorating. Urbanization over large areas, and more specifically along the Puget Sound shoreline, is suspected as a principal agent of deterioration. A statistical correlation between shoreline development and absence of herring spawn is suggestive of such a role, but the same methodology also indicated a weaker role for structural modification and with the opposite effect—an increase in herring spawn—which no one would have predicted or can explain. Counter intuitive model results, combined with knowledge gained from process-based studies and site assessments, indicate the need to investigate chains of ecological effect systematically. To understand and then manage the ecological health of Puget Sound, a Sound-wide model is necessary that incorporates multiple effects. Collection and analysis of sufficient data to develop such a model may be a decades-long effort, as has been the case for developing comprehensive knowledge of Puget Sound's counterparts Chesapeake Bay on the Atlantic coast, and San Francisco Bay, California.

References Cited

Cole, M.L., Valiela, I., Kroeger, K.D., Tomasky, G.L., Cebrian, J., Wigand, C., McKinney, R.A., Grady, S.P., and Carvalho da Silva, M.H., 2004, Assessment of a $\delta^{15}N$ isotopic method to indicate anthropogenic eutrophication in aquatic ecosystems: Journal of Environmental Quality, v. 33, p. 124-132.

Douben, P.E.T., 2003, PAHs—An ecotoxicological perspective: Chichester, England, John Wiley & Sons Ltd, 392 p.

Dougherty, J.A., Swarzenski, P.W., Dinicola, R.S., and Reinhard, M., 2010, Occurrence of pharmaceutical and personal care product residues in surface water and groundwater around Liberty Bay, Puget Sound, Washington: Journal of Environmental Quality, v. 39, no. 4, p. 1173-1180.

Driscoll, C.T., Whitall, D., Aber, J., Boyer, E., Castro, M., Cronan, C., Goodale, C.L., Groffman, P., Hopkinson, C., Lambert, K., Lawrence, G., and Ollinger, S., 2003, Nitrogen pollution in the northeastern United States—Sources, effects, and managements options: BioScience, v. 53, no. 4, p. 347-374.

Edwards, V., Icely, J., Newton, A., and Webster, R., 2005, The yield of chlorophyll from nitrogen—A comparison between the shallow Ria Formosa lagoon and the deep oceanic conditions at Sagres along the southern coast of Portugal: Estuarine, Coastal, and Shelf Science, v. 62, p. 391-403.

Eisler, R., 1985, Cadmium hazards to fish, wildlife, and invertebrates—A synoptic review: U.S. Fish and Wildlife Service Biological Report 85 (1.2), Contaminant Hazards Report No. 2, p. 46.

Ford, J., 1989, The effects of chemical stress on aquatic species composition and community structure, in Levin, S.A., ed., Ecotoxicology: New York, Springer-Verlag, p. 99–144.

Gowen, B.J., Tett, P., and Jones, K.J., 1992, Predicting marine eutrophication—The yield of chlorophyll from nitrogen in Scottish coastal waters: Marine Ecology Progress Series, v. 85, p. 153-161.

Khetan, S.K., and Collins, T.J., 2007, Human pharmaceuticals in the aquatic environment—A challenge to green chemistry: Chemical Reviews, v. 107, p. 2319-2364.

Kitsap, 2006, Liberty Bay/Miller Bay watershed 2006 water quality monitoring report: Kitsap County Health District, 14 p.

Lake, J.L., McKinney, R.A., Osterman, F.A., Pruell, R.J., Kiddon, J., Ryba, S.A., and Libby, A.D., 2001, Stable nitrogen isotopes as indicators of anthropogenic activities in small freshwater systems: Canadian Journal of Fisheries and Aquatic Sciences, v. 58, p. 870-878.

MacDonald, K., Simpson, D., Paulson, B., Cox, J., and Gendron, J., 1994, Shoreline armoring effects on physical coastal processes in Puget Sound, Washington: Olympia, Wash., Washington Department of Ecology Shorelands and Water Resources Program, 171 p.

McClelland, J.W., and Valiela, I., 1998, Linking nitrogen in estuarine producers to land-derived sources: Limnology and Oceanography, v. 43, no. 4, p. 577-585.

McKinney, R.A., Lake, J.L., Allen, M., and Ryba, S.A., 1999, Spatial variability in mussels used to assess base level nitrogen isotope ratio in freshwater ecosystems: Hydrobiologia, v. 412, p. 17-24.

Newton, J.A., Anderson, S.L., van Voorhis, K., Maloy, C., and Siegel, E., 2002, Washington State marine water quality in 1998 through 2000: Washington State Department of Ecology, Environmental Assessment Program Publication 02-03-056, 111 p.

Oczkowski, A., Nixon, S., Henry, K., DiMilla, P., Pilson, M., Granger, S., Buckley, B., Thornber, C., McKinney, R.A., and Chavez, J., 2008, Distribution and trophic importance of anthropogenic nitrogen in Narragansett Bay—An assessment using stable isotopes: Estuaries and Coasts, v. 31, p. 53-69.

Odum, E.P., 1985, Trends expected in stressed ecosystems: BioScience, v. 35, p. 419–422.

Thom, R.M., Shreffer, D.K., and MacDonald, K., 1994, Shoreline armoring effects on coastal ecology and biological resources in Puget Sound, Washington: Olympia, Wash., Washington Department of Ecology Shorelands and Water Resources Program, 102 p.

Washington State Department of Natural Resources, 2010, Available GIS data—Washington ShoreZone inventory: Washington State Department of Natural Resources database, accessed October 18, 2010, at http://fortress.wa.gov/dnr/app1/dataweb/dmmatrix.html.

Woodwell, G.M., 1983, The blue planet—Of wholes and parts and man, in Mooney, H.A., and Godron, M., eds., Disturbance and ecosystems: New York, Springer-Verlag, p. 1-10.

Suggested Citation

Takesue, R.K., Dinicola, R.S., Lacy, J.R., Liedtke, T.L., Rondorf, D.W., Smith, C.D., and Watts, R.D., 2011, Synthesis, chap. 8 of Takesue, R.K., ed., Hydrography of and biogeochemical inputs to Liberty Bay, a small urban embayment in Puget Sound, Washington: U.S. Geological Survey Scientific Investigations Report 2011–5152, p. 93-98.